AF388896

Spider Ecology and Behaviour

Spider Ecology and Behaviour

Editors

Thomas Hesselberg
Dumas Gálvez

MDPI • Basel • Beijing • Wuhan • Barcelona • Belgrade • Manchester • Tokyo • Cluj • Tianjin

Editors
Thomas Hesselberg
University of Oxford
UK

Dumas Gálvez
University of Panama
Panama

Editorial Office
MDPI
St. Alban-Anlage 66
4052 Basel, Switzerland

This is a reprint of articles from the Special Issue published online in the open access journal *Insects* (ISSN 2075-4450) (available at: https://www.mdpi.com/journal/insects/special_issues/spider_ecology_behaviour).

For citation purposes, cite each article independently as indicated on the article page online and as indicated below:

LastName, A.A.; LastName, B.B.; LastName, C.C. Article Title. *Journal Name* **Year**, *Volume Number*, Page Range.

ISBN 978-3-0365-7394-6 (Hbk)
ISBN 978-3-0365-7395-3 (PDF)

Cover image courtesy of Adrian Infernus

Contents

About the Editors

Thomas Hesselberg

Hesselberg, Thomas, PhD, Lecturer, Oxford University Department for Continuing Education Thomas is the director of studies for biological sciences and the course director for the online postgraduate certificate in ecological survey techniques at the Oxford University Department for Continuing Education. In addition, he is a college lecturer at St. Anne's College, University of Oxford, and a senior research associate at the Jersey International Centre of Advanced Studies. He has been researching spiders for more than 20 years focussing on the behaviour of orb spiders in temperate, tropical and subterranean environments. He is a section editorial board member of *Insects* and a fellow of the Royal Entomological Society.

Dumas Gálvez

Dumas Gálvez, PhD, Staff Scientist at the Coiba Scientific Station, and Lecturer at the University of Panama, Central American Program of Entomology. Dumas teaches the courses of Insect Ecology, Insect Taxonomy and Social Insects at the University of Panama. He carries out research in predator–prey interactions, using arachnids as models, as well as working with other invertebrates and vertebrates in the tropics. He is an editorial board member of *Insects* and a member of the Animal Behavior Society.

Preface to "Spider Ecology and Behaviour"

Spiders are the only large arthropod group that uses silk throughout their life; therefore, much effort has been focused on the interactions between ecology, silk, and foraging behaviour. However, it is also becoming clear that spiders can more generally act as model organisms for studies on sexual selection, invertebrate cognition, and animal communication and signalling, as well as their potential role in biological control. Similarly, new promising research directions are being developed that explore plant–spider and spider–microbe interactions.

This Special Issue broadly addresses studies on Spider Ecology and Behaviour across all relevant disciplines, including applied areas, and includes reviews, research articles, and brief reports.

Thomas Hesselberg and Dumas Gálvez
Editors

Editorial

Spider Ecology and Behaviour—Spiders as Model Organisms

Thomas Hesselberg [1,2,*] and Dumas Gálvez [3,4,5]

1 Department for Continuing Education, University of Oxford, Oxford OX1 2JA, UK
2 Department of Biology, University of Oxford, Oxford OX1 3PJ, UK
3 Coiba Scientific Station, Panama City 0843-01853, Panama
4 Programa Centroamericano de Maestría en Entomología, Universidad de Panamá, Panama City 0824, Panama
5 Smithsonian Tropical Research Institute, Panama City P.O. Box 0843-03092, Panama
* Correspondence: thomas.hesselberg@biology.ox.ac.uk

Citation: Hesselberg, T.;
Gálvez, D. Spider Ecology and
Behaviour—Spiders as Model
Organisms. *Insects* **2023**, *14*, 330.
https://doi.org/10.3390/
insects14040330

Received: 21 March 2023
Accepted: 27 March 2023
Published: 28 March 2023

1. Introduction

Spiders are versatile and ubiquitous generalist predators that can be found in all terrestrial ecosystems except for Antarctica. Therefore, it is, perhaps, unsurprising that they have been studied fairly extensively within many of the subdisciplines that make up ecology and animal behaviour. In ecology, they are prominently featured in studies, particularly on dispersal and biogeography [1,2], due to their unique ability for long-distance dispersal via ballooning [3], and in studies on niche separation [4,5]. In behavioural studies, they are model organisms for studies in animal communication and signalling [6], foraging behaviour [7], mating behaviour and animal contests [8,9], and cognition [10,11], while web-building spiders, in particular, are also used extensively in studies on construction behaviour and behavioural flexibility [12,13]. More recently, it has also become evident that spiders, in addition to their intrinsic interest as fascinating and, illogically, feared animals, likely due to press misinformation [14], may be of direct benefit to society, a field we can call applied arachnology similar to, or a subset of, the more established field of applied entomology. Areas of particular interest to behaviour and ecology include their role as enemies of natural pests [15,16], their webs as indicators of pollution [17,18], and their significant potential in biomimetics, which is the inspiration, abstraction, and application of evolved processes or traits in biological organisms to our technology [19]. The biomimetics potential of spider behaviour includes biologically inspired locomotion and robotics [20,21], and using the spider web for inspiration for sensors [22] and light weight composite structures [23].

This Special Issue reflects the diverse range of topics within ecology and behaviour that can be fruitfully studied using spiders as model organisms. Below, we give a brief overview of the papers featured in the Special Issue in the context of applied and basic research and highlighting two papers that evaluate and develop new methods for studying their behaviour.

2. Applied Arachnology

One of the most well-studied areas of applied arachnology is undoubtedly arachnids' potential role as natural enemies of agricultural pests. Traditionally, the most focus has been on mites, which can act both as pests [24] and as natural pest controllers [25]. However, the role of spiders as important regulators of pest species in agroecosystems, in combination with parasitic wasps and other insect predators, is becoming more established [16]. In the present Special Issue, Thomas Roberts-McEwen and colleagues [26] show that a group-living araneid, the tropical tent-web spider, *Cyrtophora citricola*, might have hitherto overlooked potential in controlling the tomato leafminer (*Tuta absoluta*), a major pest on tomato plants worldwide. Choice experiments demonstrated that the tent-web spider had nearly similar capture efficiency between tomato leafminers and mutant flight-less *Drosophila*, and had a far higher capture success rate than against the larger black soldier flies. This, combined with observational data from Southern Spain on web sizes in different seasons, suggests that the tent-web spider could potentially be a successful biological control agent of the tomato leafminer in the tomato planting and growing season assuming that high parasitic wasp infection of spider eggs can be controlled [26].

3. Basic Research on Spider Ecology and Behaviour

Spider foraging strategies, and thus, to some extent, most aspects of their ecology and behaviour, can be split into either active roaming spiders that do not build a web (cursorial spiders) or sit-and-wait web-building spiders. The Special Issue includes two papers that focus on cursorial spiders, and five papers with a main focus on web-building spiders.

3.1. Cursorial Spiders

Maria Trabalon [27] looked at spider breeding welfare and compared the body mass and locomotory and exploratory behaviour of the wolf spider, *Pardosa saltans*, between spiders, immediately after they were caught, wild-caught spiders after being kept 15 days in the laboratory and laboratory-reared spiders. The results showed that while laboratory rearing increased body mass, it reduced behavioural activities, although this reduction could be mitigated by providing litter to the rearing chambers. Marzena Stańska and Tomasz Stański [28] compared assemblages, including both cursorial and web-building spiders inhabiting the optimal, terminal/decay, and regeneration phases of a primeval forest in Poland. Interestingly, the study suggests that the highest species diversity is found in the terminal/decay phase, possibly due to more niches in that phase, while the regeneration phase had the lowest.

3.2. Web-Building Spiders

We start this section on web-building spiders with a very interesting group of araneid spiders that have lost the ability to build full orb webs—the bolas spiders, which, instead of a web construct, they use a single thread as a lasso to catch moths attracted by the pheromones emitted by the spider [29]. Candido Dias, Jr and colleagues looked closely at the biomechanics behind this fascinating prey capture behaviour. In the first paper [30], Candido Dias, Jr and John Roff showed, using high-speed cameras, that the South African grassland bolas spider, *Cladomelea akermani*, actively spins, not only the bolas, but also its body. They were able to show with computational fluid dynamics models that this spinning likely has the function of further spreading the emitted pheromones in open habitats. In the second paper [31], Candido Dias, Jr. and John Long, Jr. analysed the prey capture behaviour of the American bolas spider, *Mastophora hutchinsoni*, by calculating the kinematics of both spider and moth based on high-speed recordings to model the physical properties of the bolas during prey capture. Their model showed that the material properties of the glue in the bolas of *M. hutchinsoni* are different to that of previously studied bolas spiders.

The Special Issue also features two brief reports on spiders and invasive species. In the first [32], Arty Schronce and Andrew Davies studied an interesting interaction in the US between the invasive Joro spider, *Trichonephila clavate*, and the native northern cardinal, where the bird perches on top of the very strong orb web and steals prey items directly from the web. In the second [33], El Ellsworth and colleagues looked at how invasive plants and other management strategies impact spider communities (predominantly web-building spiders) in five parks in the greater Memphis area in the US. The study showed that invasive plants can serve as a useful habitat for native spiders as exemplified by the native humpbacked orb-weaver, *Eustala anastera*, being found exclusively on the invasive Chinese privet.

Lastly, in a review, Thomas Hesselberg and colleagues [34] looked more broadly at the associations between web-building spiders and specific host plant species, including a brief overview of cursorial spider-plant associations. The study confirms that associations between spiders and plants are rare, but also found two promising candidates for further studies. The Australian linyphiid *Laetesia raveni* is exclusively reported from two thorny plant species, and two species of Central American araneids in the genus *Eustala* are tightly associated with ant-protected acacia trees.

4. New Methodologies for Studying Spider Ecology and Behaviour

Novel methodologies or approaches to conducting research is often a major driver for important research breakthrough, and the field of spider ecology and behaviour is no exception. The field is particularly diverse in its methodology ranging from low-tech, cheap, and simple experimental approaches [35] to high-tech computational or experimental approaches [36]. The Special Issue includes two papers from each of these extremes. In the latter category, Nathan Justus and colleagues [37] developed a clever high-tech integrated system of combining stereo vision and video vibrometry to automatically gather 3D vibrational information to study signal propagation in spider webs, which they successfully validated using laser vibrometry in webs of black widows (*Latrodectus hesperus*). In the former category, Mollie Davies and Thomas Hesselberg [38] reviewed and updated a cheap and easy, old technique of studying behaviour of orb web-building spiders in the field using a tuning fork. They showed that while high-frequency tuning forks (440 Hz) mostly elicited prey capture behaviour in the tetragnathid spider *Metellina segmentata*, a lower-frequency tuning fork (256 Hz) tended to elicit escape behaviour.

5. Conclusions

Hopefully, it will be clear from the overview given above that the Special Issue covers a wide range of topical areas of spider research within the fields of ecology and behaviour. We hope that the ten papers included in the issue will contribute to further stimulate research in ecology and behaviour using spiders as model organisms.

Author Contributions: Conceptualization, T.H. and D.G.; writing—original draft preparation, T.H.; writing—review and editing, D.G. All authors have read and agreed to the published version of the manuscript.

Funding: The authors gratefully acknowledge funding from the Panamanian government (SENACYT grant FID22-034), Sistema Nacional de Investigación, and a Royal Society International Exchanges grant (IES\R3\213007).

Conflicts of Interest: The authors declare no conflict of interest.

References

1. Bronte, D. Case-study II—spiders as a model in dispersal ecology and evolution. In *Dispersal Ecology and Evolution*; Clobert, J., Baguette, M., Benton, T.G., Bullock, J.M., Eds.; Oxford University Press: Oxford, UK, 2012.
2. Gillespie, R.G. Biogeography of spiders on remote oceanic islands of the Pacific: Archipelagoes as stepping stones? *J. Biogeogr.* **2002**, *29*, 655–662. [CrossRef]
3. Weyman, G.S. A review of the possible causative factors and significance of ballooning in spiders. *Ethol. Ecol. Evol.* **1993**, *5*, 279–291. [CrossRef]
4. Brown, K.M. Foraging ecology and niche partitioning in orb-weaving spiders. *Oecologia* **1981**, *50*, 380–385. [CrossRef] [PubMed]
5. Sanders, D.; Vogel, E.; Knop, E. Individual and species-specific traits explain niche size and functional role in spiders as generalist predators. *J. Anim. Ecol.* **2015**, *84*, 134–142. [CrossRef] [PubMed]
6. Oxford, G.S.; Gillespie, R.G. Evolution and ecology of spider colouration. *Annu. Rev. Entomol.* **1998**, *43*, 619–643. [CrossRef]
7. Uetz, G. Foraging strategies of spiders. *Trends Ecol. Evol.* **1992**, *7*, 155–159. [CrossRef]
8. Arnott, G.; Elwood, R.W. Assessment of fighting ability in animal contests. *Anim. Behav.* **2009**, *77*, 991–1004. [CrossRef]
9. Elias, D.O.; Kasumovic, M.M.; Punzalan, D.; Andrade, M.C.B.; Mason, A.C. Assessment during aggressive contests between male jumping spiders. *Anim. Behav.* **2008**, *76*, 901–910. [CrossRef]
10. Jackson, R.R.; Cross, F.R. Spider Cognition. In *Spider Physiology and Behaviour*; Casas, J., Ed.; Academic Press: Cambridge, MA, USA, 2011; Volume 41, pp. 115–174.
11. Japyassú, H.F.; Laland, K.N. Extended spider cognition. *Anim. Cogn.* **2017**, *20*, 375–395. [CrossRef] [PubMed]
12. Eberhard, W.G. *Spider Webs—Behaviour, Function and Evolution*; University of Chicago Press: Chicago, IL, USA, 2020.
13. Hesselberg, T. Exploration behaviour and behavioural flexibility in orb-web spiders: A review. *Curr. Zool.* **2015**, *61*, 313–327. [CrossRef]

14. Mammola, S.; Malumbres-Olarte, J.; Arabesky, V.; Barrales-Alcalá, D.A.; Barrion-Dupo, A.L.; Benamú, M.A.; Bird, T.L.; Bogomolova, B.M.; Cardoso, P.; Chatzaki, M.; et al. The global spread of misinformation on spiders. *Curr. Biol.* **2022**, *32*, R871–R873. [CrossRef]

15. Sunderland, K. Mechanisms Underlying the Effects of Spiders on Pest Populations. *J. Arachnol.* **1999**, *27*, 308–316.

16. Michalko, R.; Pekár, S.; Entling, M.H. An updated perspective on spiders as generalist predators in biological control. *Oecologia* **2019**, *189*, 21–36. [CrossRef]

17. Samu, F.; Vollrath, F. Spider orb web as bioassay for pesticide side effects. *Entomol. Exp. Appl.* **1992**, *62*, 117–124. [CrossRef]

18. Stojanowska, A.; Zeynalli, F.; Wróbel, M.; Rybak, J. The use of spider webs in the monitoring of air quality—A review. *Integr. Environ. Assess. Manag.* **2023**, *19*, 32–44. [CrossRef] [PubMed]

19. Lenau, T.A.; Metze, A.-L.; Hesselberg, T. Paradigms for biologically inspired design. In Proceedings of the SPIE Smart Structures and Materials + Nondestructive Evaluation and Health Monitoring, Denver, CO, USA, 5–6 March 2018; Volume VIII, p. 1059302.

20. Vidoni, R.; Gasparetto, A. Efficient force distribution and leg posture for a bio-inspired spider robot. *Robot. Auton. Syst.* **2011**, *59*, 142–150. [CrossRef]

21. Vollrath, F.; Krink, T. Spider webs inspiring soft robotics. *J. R. Soc. Interface* **2020**, *17*, 20200569. [CrossRef] [PubMed]

22. Zhou, J.; Lai, J.; Menda, G.; Stafstrom, J.A.; Miles, C.I.; Hoy, R.H.; Miles, R.N. Outsourced hearing in an orb-weaving spider that uses its web as an auditory sensor. *Proc. Natl. Acad. Sci. USA* **2022**, *119*, e2122789119. [CrossRef]

23. Regassa, Y.; Lemu, H.G.; Sibarizuh, B.; Rahimeto, S. Studies on the Geometrical Design of Spider Webs for Reinforced Composite Structures. *J. Compos. Sci.* **2021**, *5*, 57. [CrossRef]

24. Takafuji, A.; Ozawa, A.; Nemoto, H.; Gotoh, T. Spider Mites of Japan: Their Biology and Control. *Exp. Appl. Acaraol.* **2000**, *24*, 319–335. [CrossRef] [PubMed]

25. McMurtry, J.A.; Croft, C.A. Life-styles of phytoseiid mites and their roles in biological control. *Annu. Rev.* **1997**, *42*, 291–321. [CrossRef] [PubMed]

26. Roberts-McEwen, T.A.; Deutsch, E.K.; Mowery, M.A.; Grinsted, L. Group-Living Spider *Cyrtophora citricola* as a Potential Novel Biological Control Agent of the Tomato Pest *Tuta absoluta*. *Insects* **2023**, *14*, 34. [CrossRef]

27. Trabalon, M. Effects of Wolf Spiders' Captive Environment on Their Locomotor and Exploratory Behaviours. *Insects* **2022**, *13*, 135. [CrossRef]

28. Stańska, M.; Stański, T. Spider Assemblages of Tree Trunks and Tree Branches in Three Developmental Phases of Primeval Oak–Lime–Hornbeam Forest in the Białowieża National Park. *Insects* **2022**, *13*, 1115. [CrossRef]

29. Yeargan, K.N. Biology of bolas spiders. *Annu. Rev. Entomol.* **1994**, *39*, 81–99. [CrossRef]

30. Diaz, C., Jr.; Roff, J. Mechanics of the Prey Capture Technique of the South African Grassland Bolas Spider, *Cladomelea akermani*. *Insects* **2022**, *13*, 1118. [CrossRef] [PubMed]

31. Diaz, C., Jr.; Long, J.H., Jr. Behavior and Bioadhesives: How Bolas Spiders, *Mastophora hutchinsoni*, Catch Moths. *Insects* **2022**, *13*, 1166. [CrossRef]

32. Schronce, A.; Davis, A.K. Novel Observation: Northern Cardinal (*Cardinalis cardinalis*) Perches on an Invasive Jorō Spider (*Trichonephila clavata*) Web and Steals Food. *Insects* **2022**, *13*, 1049. [CrossRef] [PubMed]

33. Ellsworth, E.; Li, Y.; Chari, L.D.; Kron, A.; Moyo, S. Tangled in a Web: Management Type and Vegetation Shape the Occurrence of Web-Building Spiders in Protected Areas. *Insects* **2022**, *13*, 1129. [CrossRef] [PubMed]

34. Hesselberg, T.; Boyd, K.M.; Styrsky, J.D.; Gálvez, D. Host Plant Specificity in Web-Building Spiders. *Insects* **2023**, *14*, 229. [CrossRef]

35. Zschokke, S.; Herberstein, M.E. Laboratory methods for maintaining and studying web-building spiders. *J. Arachnol.* **2005**, *33*, 205–213. [CrossRef]

36. Mortimer, B.; Soler, A.; Siviour, C.R.; Zaera, R.; Vollrath, F. Tuning the instrument: Sonic properties in the spider's web. *J. R. Soc. Interface* **2016**, *13*, 20160341. [CrossRef] [PubMed]

37. Justus, N.; Krugner, R.; Hatton, R.L. Validation of a Novel Stereo Vibrometry Technique for Spiderweb Signal Analysis. *Insects* **2022**, *13*, 310. [CrossRef]

38. Davies, M.S.; Hesselberg, T. The Use of Tuning Forks for Studying Behavioural Responses in Orb Web Spiders. *Insects* **2022**, *13*, 370. [CrossRef]

Article

Group-Living Spider *Cyrtophora citricola* as a Potential Novel Biological Control Agent of the Tomato Pest *Tuta absoluta*

Thomas A. Roberts-McEwen [1,*], Ella K. Deutsch [2], Monica A. Mowery [3] and Lena Grinsted [1]

[1] School of Biological Sciences, University of Portsmouth, King Henry Building, King Henry 1 Street, Portsmouth PO1 2DY, UK
[2] School of Life Sciences, University of Nottingham, University Park, Nottingham NG7 2RD, UK
[3] Jacob Blaustein Institutes for Desert Research, Mitrani Department of Desert Ecology, Ben-Gurion University of the Negev, Midreshet Ben-Gurion 8499000, Israel
[*] Correspondence: thomas.roberts-mcewen@myport.ac.uk or t.robertsmcewen@gmail.com

Simple Summary: The tomato leafminer, *Tuta absoluta*, is a devastating pest moth of commercially important crops like tomato and potato. This moth has developed resistance to insecticides; therefore, novel approaches, like using natural predators, are needed to combat infestations. We explored the use of tropical tent web spiders, *Cyrtophora citricola*, as biological control agents, as these spiders live in groups and are not cannibalistic, and thus, create large, predator-dense webs. Furthermore, their global range overlaps with regions of moth infestations. In lab settings, we introduced different prey types to small colonies of spiders of varying body sizes and found that spiders were equally efficient at capturing pest moths and easily-caught fruit flies (*Drosophila hydei*). Larger spiders built larger webs and were better at catching prey. Spiders from southern Spain were large enough to capture pest moths during the tomato growing season, but >50% of spider egg sacs were attacked by egg predatory wasps (*Philolema palanichamyi*). *Cyrtophora citricola* spiders, therefore, have the potential to be an effective biological control agent of flying insect pests, at least after growing to medium-sized juveniles, and if wasp infections are controlled, forming part of integrated pest management to defend against pest infestations in the future.

Citation: Roberts-McEwen, T.A.; Deutsch, E.K.; Mowery, M.A.; Grinsted, L. Group-Living Spider *Cyrtophora citricola* as a Potential Novel Biological Control Agent of the Tomato Pest *Tuta absoluta*. *Insects* **2023**, *14*, 34. https://doi.org/10.3390/insects14010034

Academic Editors: Thomas Hesselberg and Dumas Gálvez

Received: 31 October 2022
Revised: 9 December 2022
Accepted: 23 December 2022
Published: 30 December 2022

Abstract: Group-living spiders may be uniquely suited for controlling flying insect pests, as their high tolerance for conspecifics and low levels of cannibalism result in large, predator dense capture webs. In laboratory settings, we tested the ability of the facultatively communal spider, *Cyrtophora citricola*, to control the tomato leafminer, *Tuta absoluta*; a major pest of tomato crops worldwide. We tested whether prey capture success was affected by spider body size, and whether prey capture differed among *T. absoluta*, flightless fruit flies (*Drosophila hydei*), and larger black soldier flies (*Hermetia illucens*). We found that larger spiders generally caught more prey, and that prey capture success was similar for *T. absoluta* and easily caught fruit flies, while black soldier flies were rarely caught. We further investigated the seasonal variations in web sizes in southern Spain, and found that pest control would be most effective in the tomato planting and growing season. Finally, we show that *C. citricola* in Spain have >50% infection rates of an egg predatory wasp, *Philolema palanichamyi*, which may need controlling to maintain pest control efficacy. These results suggest that using *C. citricola* as a biological control agent in an integrated pest management system could potentially facilitate a reduction of pesticide reliance in the future.

Keywords: sociality; communal; colonial spiders; predator-prey; food security; pesticide resistance; sustainable agriculture

1. Introduction

Climate change due to human overpopulation and fossil fuel dependence is facilitating the spread of invasive pest species of agricultural crops by expanding their habitable environment ranges [1]. Increasing interconnectedness of the global food chain also allows for the anthropogenic introduction of potentially devastating agricultural pests, increasing pesticide reliance worldwide [1]. The tomato leafminer, *Tuta absoluta* (Meyrick, 1917) (Lepidoptera: Gelechiidae), is a species that has undergone rapid range expansion, reaching near-global ubiquity [2–4]. *Tuta absoluta* is a species of neotropical, oligophagous moths of solanaceous crops, with a preference for tomato [5]. The species also associates with a number of other host plants, many of which are agricultural crops, such as potatoes, bell and chilli peppers, and aubergine [5]. Crop damage is caused by larval feeding, which affects all epigeal plant parts, most notably the leaves [5,6]. Larvae burrow to consume the mesophyll layer, reducing photosynthetic surface area and resulting in diminished plant growth and fruit yield (Figure 1a) [6]. Larvae can also directly attack the tomato fruit, causing aesthetic damage and rendering the crop unmarketable [4]. Larvae live within the leaf until they pupate, making them difficult to control during juvenile life stages, due to predators lacking the capability to effectively target them within the leaf [6,7]. Here, we explore the potential for the use of the group-living tropical tent web spider, *Cyrtophora citricola* (Forskål, 1775) (Araneae: Araneidae) as a biological control agent of the adult, flying moth. The geographical distributions of *C. citricola* and *T. absoluta* overlap in large parts of the world, making this web-building spider a suitable candidate for biological pest control.

Tuta absoluta has infested 60% of global tomato-cultivated land [2], and can cause 80–100% yield reduction in both open-field and protected cultivations if left untreated [5]. As tomato is among the most cultivated and consumed vegetable crops worldwide [5], finding suitable control strategies to reduce the ubiquity and feeding voracity of *T. absoluta* is of increasing importance [3,8]. *Tuta absoluta* infestations are causing yield loss and, therefore, economic detriment that disproportionately affects low- and lower-middle-income countries (LMICs) [9,10]. Chemical insecticides have historically been favoured as mitigators of localised damage caused by numerous phytophagous insect species [11]. However *T. absoluta* has developed resistance to many commonly used synthetic insecticides, ultimately resulting in difficulty controlling infestations [2,4,12,13]. Furthermore, pesticide dependency in LMICs has resulted in a wide range of detrimental outcomes for both human and environmental health [10]. As the workforce in LMICs moves away from farming and toward industrialised society, domestic food is increasingly produced by fewer, often educationally disadvantaged individuals, and LMICs with long growing seasons resort to increasing non-traditional crop export to temperate zones to earn valuable foreign currency [10,14]. These socio-economic and agricultural shifts are not currently possible without increased crop yield, facilitated by the use of chemical pesticides, many of which are illicit, homemade mixes that are sold more affordably than those that are regulated [14]. Furthermore, these pesticides are often used by farmers with little means of procuring appropriate personal protective equipment (PPE), or little power or willingness to ensure its use within the workforce [14], resulting in frequent incidences of poisoning [15,16].

Currently, chemical insecticides remain the most widely used method of controlling *T. absoluta*, despite their inefficiency and danger to humans and the environment [13]. It is, therefore, important to explore whether natural predators of agricultural pest species can be used to negate the detrimental effects of herbivore infestation, with the aim of reducing the reliance on mass distribution of toxic chemical pesticides [15,16]. Research into finding appropriate biological control agents against *T. absoluta* is ongoing. Due to the ubiquity of the pest, it is likely that multiple biological control species will be required to meet the needs of the diverse ecological systems and climates it has invaded. Various approaches to *T. absoluta* population management have been tested, including the use of predators, parasitoids, and entomopathogens [17]. It is likely that a combination of these methods will prove most effective as components of an integrated pest-management system [18]; however, the search for a highly effective combination of natural predators is ongoing [4]. Furthermore, most current biological control methods for *T. absoluta* infestation rely on controlling the pest at its larval stage [19]. Methods currently used for targeting exclusively larval *T. absoluta* instars include, but are not limited to: the use of bacterial toxins produced by *Bacillus thuringiensis* (Berliner, 1915) (Bacillales: Bacillaceae) [20]; granulovirus isolates from *Phthorimaea operculella* (Zeller, 1873) (Lepidoptera: Gelechiidae) that delays *T. absoluta* larval growth and reduces pupation [21]; larval parasitoids such as *Dolichogenidea (=Apanteles) gelechiidivoris* (Marsh, 1975) (Hymenoptera: Braconidae) [22]; entomopathogenic fungi, such as *Beauveria bassiana* (Bals.-Criv., Vuill., 1912) (Hypocreales: Cordycipitaceae) and *Metarhizium anisopliae* (Metschn., Sorokīn, 1883) (Ascomycota: Hypocreales) [23]; and entomopathogenic nematodes, such as *Heterorhabditis bacteriophora* (Poinar, 1976) (Nematoda: Heterorhabditidae) and *Steinernema carpocapsae* (Weiser, 1955) (Rhabditida: Steinernematidae) [24]. Due to the short life cycle of *T. absoluta*, as well as their overlapping generations [2], simultaneous removal of individuals at all instars must be achieved to produce an integrated pest-management approach effective enough to prevent *T. absoluta* reinfestation.

Figure 1. (**a**) *Tuta absoluta* larva (circled in red) feeding on the mesophyll layer of a tomato leaf. Picture taken in laboratory conditions; (**b**) an adult *C. citricola* individual in natural field settings in southern Spain, with four egg sacs; (**c**) *C. citricola* colony with visible individual horizontal web sheets on *Opuntia* sp. cactus in natural field settings in southern Spain; (**d**) spiderlings and egg sac in a 40 mL falcon tube; (**e**) colony of spiders on wire netting in a large sized mesh enclosure; (**f**) small, medium, and large sized mesh enclosures containing wire web supports for *C. citricola* colonies. Similar mesh enclosures were also used to rear *T. absoluta* moths on tomato plants. (**a,d–f**) were photographed in the laboratory in Portsmouth, UK. Photos: (**c**) LG; (**a,b,d–f**) TARM.

Web-building spiders are key predators of flying insects [18]. One particularly promising yet unexplored potential biological control agent of the adult instar of *T. absoluta* is *C. citricola*, a species of facultatively group-living orb-weaving spider (Figure 1b) [25,26]. *Cyrtophora citricola* occurs in Mediterranean Europe, Africa, Asia, and the Middle East [25,27], all of which are regions that contain LMICs suffering from *T. absoluta* invasion [10,19,28], highlighting the geographical suitability of *C. citricola* as a biological control agent of *T. absoluta*. Furthermore, the group-living tendencies of *C. citricola*, with their high levels of conspecific tolerance and low levels of cannibalism, can result in high densities of predators [26,29,30]. *Cyrtophora citricola* produce non-adhesive, horizontal sheet webs, which are defended territorially from conspecifics despite high tolerance of colony members occupying adjacent webs (Figure 1c,e) [25,31,32]. Spiders commonly attach their individual webs together to form large colonies, and the connecting threads in *C. citricola* colonies are used communally (Figure 1c,e) [33,34]. Colony geometry facilitates the exploitation of the 'ricochet effect', wherein prey items that escape the web of one individual may fall into the web of an adjacent conspecific in the colony [35]. The importance of the unique construction of the three-dimensional capture web structure for enhanced prey capture capability is further described in Su and Buehler (2020), who suggest that connecting threads also play a role in filtering out prey of low impact velocity, and protecting the individual residing at the centre of the web [36]. Two major components that contribute to the robustness and fast repairability of *C. citricola* webs are: (a) that web silk displays non-linear behaviour, wherein it may soften or stiffen at a molecular level as a response to strain, and (b), that tension in the main load-bearing strand can be released to multiple other strands in the event of it breaking, preventing web collapse [36]. Therefore, in the event of damage being done to *C. citricola* web structures, biological control capability can be effectively maintained. Their potential for effective use as biological control agents is therefore greater than that of more aggressive, solitary spiders that are prone to cannibalism [37–39]. Past studies on the use of spiders for pest control predominantly focus on non-web-building, solitary species, due to their propensity to capture prey from the crop surface, as well as their ability to consume less motile prey arthropod instars [7,40,41]. However, spiders that can form groups of hundreds, or even thousands, of interconnected webs can provide large surface areas of capture webs capable of intercepting high frequencies of airborne arthropods [37,42]. *Cyrtophora citricola* colonies also provide a substrate for other spider species, such as kleptoparasitic *Argyrodes* spp. (Araneae: Theridiidae), and *Holocnemus pluchei* (Araneae: Pholcidae) [33], further increasing predator density and, therefore, potentially increasing pest insect capture capability within colonies.

In this study, we ask whether *C. citricola* has the potential to act as a biological control agent of *T. absoluta*. We use southern Spain as a case study, where *C. citricola* has a strong association with prickly pear cactus (*Opuntia* spp.) that has historically been used as fences and field borders around agricultural fields [42,43]. Therefore, spider colonies are commonly found in association with field-grown crops in this region, with a potential for providing biological pest control [44]. First, we record the capture rate of *T. absoluta* by small spider colonies in lab settings, and test the effect of spider body size (and therefore spider web size) on capture success. We compare the capture rates of *T. absoluta* with that of easily caught flightless fruit flies (*Drosophila hydei*) as a control. Next, we investigate the seasonal variations in spider web sizes in southern Spain and relate that, and the potential for *T. absoluta* control, to the tomato-growing season. Finally, we consider possible inhibitors of *C. citricola* prey capture efficiency, with a focus on the egg predator *Philolema palanichamyi* (Narendran, 1984) (Hymenoptera: Eurytomidae).

This paper aims to: (a) compare the ability of *C. citricola* to capture three different prey items: the tomato leafminer, *T. absoluta*; the easily-caught and similarly sized flightless fruit flies, *D. hydei*; and the much larger black soldier flies, *Hermetia illucens*; and whether a prey type preference was exhibited; (b) test the effect of *C. citricola* body size on prey capture capability and prey size preference; (c) assess whether seasonal changes in web size may affect prey capture efficacy and, therefore, biological control potential of *C. citricola* in southern Spain; and (d) estimate the potential for egg predatory wasps, *P. palanichamyi*, to negatively affect *C. citricola* pest control capability. Spiders used in the laboratory-based experiments were reared from eggs in the laboratory, whereas seasonal web size and wasp infection data were collected from wild populations.

2. Materials and Methods

2.1. Collecting and Rearing C. citricola Spiderlings Prior to Experimental Setup

Egg sacs of *C. citricola* (N = 87) were collected in May 2021 from four sites in southern Spain, near Cádiz, coded PA (36°40′20.39″ N, 6°23′25.91″ W), CM (36°39′25.07″ N, 6°22′19.23″ W), MA (36°39′49.41″ N, 6° 5′54.14″ W), and AQ (36°37′25.00″ N, 6°11′42.96″ W). *Cyrtophora citricola* construct a string of multiple egg sacs (Figure 1b) [45]. Only the most recently constructed egg sacs (the bottom egg sacs on each string) were collected, to prevent spiderlings emerging in transit to the UK (Figure 1b). Egg sacs were reared at 22 °C in the laboratory at the University of Portsmouth in 40 mL plastic collection tubes with foam bungs (Figure 1d). Tubes were sprayed with a very fine mist of water three times per week.

After emerging and creating capture webs within the 40 mL falcon tubes, each clutch of spiderlings was fed with five to eight D. hydei once per week. As spiderlings grew, they were transferred to either: (a) one litre (1 L) plastic tubs (diameter: 12 cm, H: 15 cm) with mesh fabric lids and web supports made from 10 cm tall rectangular wire (H:10 × L:30 cm) that was rolled up along its length to provide a structure for webs of varying sizes; or (b) large 90 cm tall (W:60 × D:60 × H:90 cm) mesh enclosures with rolled up 70 cm tall (H:70 × L:100 cm) rectangular wire as web supports (Figure 1f). Food supply was increased to ten to fifteen *D. hydei* per clutch per week as spiderlings grew. All spiderlings were kept at room temperature (~22 °C) during development. No spiders reached their final moult (sexual maturity) during this time, and therefore, due to time constraints, all spiders used in the study were juveniles in varying stages of pre-adult development. Prey capture assays were conducted from September to November 2021.

2.2. Rearing T. absoluta Moths and Flies

Tomato leaves infested with *T. absoluta* larvae were first provided by a local UK tomato grower (N_{larvae}~50). Later, as our experiment continued past the UK tomato growing season, additional *T. absoluta* pupae (N_{pupae}~100) were provided by the Centre for Agriculture and Bioscience International (CABI), Ghana. Moths from both the UK and Ghana were randomly allocated amongst experimental spider colonies, ensuring no correlation between spider body sizes and moth origin. *Tuta absoluta* larvae require fresh tomato leaves for completion of their life cycle, thus, galleries containing the larvae were carefully cut from the infected leaves and placed on lab-grown tomato plants. Most larvae successfully burrowed into the mesophyll layer of the living tomato plants (Figure 1a). The larvae and tomato plants were contained in mesh enclosures of varying sizes (small: W:30.5 × D:30.5 × H:30.5 cm; medium: W:40 × D:40 × H:60 cm; large: W:60 × D:60 × H:90 cm (Figure 1f)) depending on the size of the tomato plants, with mesh holes roughly 0.5 mm in diameter. Double-sided adhesive tape was applied to the table surface surrounding each enclosure to ensure that any potential escaping larvae would be caught by sticking onto the tape.

The larvae were left in the enclosures with tomato plants to pupate until emerging as adults, when they were removed and transferred to the spider enclosures for the prey capture assays (see below). Moths were carefully caught in 40 mL tubes within the enclosure to prevent escape. The larvae and tomato plants were sprayed with water three times per week, and the tomato plants were watered once per week. After moth rearing was complete, all enclosures and plants were frozen at −6 °C for four weeks to prevent any remaining *T. absoluta* larvae, pupae, or adults from surviving.

Black soldier flies (*H. illucens*) and fruit flies (*D. hydei*) were purchased from an online pet food supplier (Livefood UK Ltd., Axbridge, UK). The *H. illucens* were received as larvae, and were transferred to small square mesh enclosures (W:30.5 × D:30.5 × H:30.5 cm) (Figure 1f) until they eclosed. The *D. hydei* were also received as larvae, and were kept in their original cultures to eclose. All flies were kept at around 22 °C in the laboratory. If too many flies were eclosing at once, and would die before being used as prey for the spiders, the culture was refrigerated at 4–5 °C to slow development.

2.3. Experimental Setup of Prey Capture Assays

We created twenty experimental spider colonies, each with a colony size of five spiders. Experimental colonies were created using spiders from 13 egg sacs from the PA site only, as this site yielded the most spiderlings. The spiderlings had grown at different rates in the lab, and we selected a total of 100 juvenile spiders of as broad a range of body sizes as possible, weighing each of them to the nearest 0.1 mg using a Sartorius B120S scale. We then temporarily placed them individually in 40 mL tubes (Figure 1d) and ranked them according to body mass (spiders ranged from 0.1 mg to 52.6 mg with a mean of 5.3 mg). Next, we placed them all into twenty colonies named A to T, each colony containing five similar-sized spiders. Due to random variations in growth rates amongst juveniles, spiders had reached different juvenile instars. The smallest five individuals comprised colony A, and the largest five individuals comprised colony T. Only female spiders were used because females are the large and communal sex in this species [37,42]. Females were identified by their lighter colouration and smaller pedipalps in comparison to male counterparts. Any juveniles that could not be sexed were excluded.

Colonies A–J were established in small mesh enclosures (W:30.5 × D:30.5 × H:30.5 cm), while Colonies K–T were established in medium mesh enclosures (W:40 × D:40 × H:60 cm) to minimise the risk of cannibalism due to crowding (Figure 1f). Rolled-up wire netting provided structural support for web building (netting in small enclosures: H:20 × L:50 cm; medium enclosures: H:40 × L:70 cm) (Figure 1e,f). The enclosures were stored adjacent to laboratory windows to receive natural light, and additional electrical room lighting was provided for 8 h per day. Any spider that died during the study was replaced by another individual in the same weight bracket from the pool of spiders from the PA site and replacements were limited to one spider per colony to minimise disturbance to group composition. In total, sixteen replacements were made.

The prey capture assays were conducted from 30 September 2021 to 22 November 2021. At the end of the 6.5-week study, 84 out of 100 spiders remained. Throughout the experimental period, spiders were sprayed with water and fed three times per week.

After the end of the experiment on 26 November 2021, spider body mass, body length, and capture web diameter were measured to assess the correlations between spider size, capture web size, and prey capture success. First, each colony was sprayed with a fine mist of water to improve web visibility. Capture web sheet diameter was then measured to the nearest 1 cm with a 30 cm ruler. Next, spiders were removed from their colony and weighed individually to nearest 0.1 mg. We measured body length from the tip of the prosoma to the bottom of the abdomen to the nearest 0.01 mm using an electronic calliper [46]. To record body length, each spider was transferred to a tray and left undisturbed until becoming still, wherein the measurement was taken with little handling to mitigate stress to the animals. All spiders were then transferred into mixed colonies and kept in the laboratory under similar conditions as described above. It was not possible to follow individual spider growth from pre- to post- experiment, so we calculated average body masses, body lengths, and web sizes per experimental colony.

2.4. Prey Capture Assays

Four prey capture treatments were implemented: (1) a control treatment introducing five flightless fruit flies per colony to represent easy-to-catch prey that was of a similar size to *T. absoluta* ($N_{\text{trials per colony}} = 5$); (2) a single *T. absoluta* per colony ($N_{\text{trials per colony}} = 3$); (3) one flightless fruit fly together with one *T. absoluta* per colony to test for prey type preference ($N_{\text{trials per colony}} = 3$); and (4) one black soldier fly per colony to represent a relatively large prey item ($N_{\text{trials per colony}} = 3$). The choice of fly species was made partly due to their accessibility from live food retailers. Average body mass for each insect was as follows (based on weighing five live specimens per species to nearest 0.1 mg): *T. absoluta*: 1.16 mg (st.dev = 0.27 mg); *D. hydei*: 2.28 mg (st.dev = 0.19 mg); *H. illucens*: 34.24 mg (st.dev = 5.47 mg). The order of capture treatment was random with respect to colony ID, and opportunistically implemented according to when prey types became available. Flies were refrigerated for five minutes to slow their movement prior to transfer to spider enclosures.

A trial consisted of placing a single insect, or several insects, according to treatment, at the bottom of a spider enclosure, to allow prey to move about and land in spider webs of their own accord, between 10 am and 12 noon. To allow enough time for insects to intercept the spider webs, colonies were left undisturbed for the following 72 h (+/−2 h). At the end of this period, the number of insects trapped in webs was counted for each colony. All insects (both live and dead) were then removed from each enclosure and replaced with fresh ones, except in the case of *T. absoluta*, where uncaptured individuals were re-used in other trials due to short supply.

2.5. Seasonal Web Size Measurement and Effects of Egg Predators

Six sites around Rota, southern Spain, coded CM, PA, NN (36°39′50.38″ N, 6°22′8.56″ W); EO (36°40′35.35″ N, 6°24′6.73″ W); WP (36°40′8.13″ N, 6°23′21.48″ W); and SN (36°38′58.15″ N, 6°22′32.79″ W) were visited roughly every 6 weeks over ten months from March 2019 to January 2020 (dates: 29/03, 06/05, 09/06, 19/08, 06/10, 29/11, 24/01). At each site, between 9 and 39 m of prickly pear cactus (*Opuntia* spp.), located along field edges, were selected for seasonal observations of spider colonies. During each trip, the horizontal web sheet diameter of all individual *C. citricola* spider webs (except from very small and hard to spot hatchling webs of just a few cm) along the stretches of cactus ($N_{\text{total web diameters}}$ = 1238) were measured to the nearest cm using a measuring tape.

Additionally, during each trip, up to three egg sac strings were collected from each of the same field sites to assess egg predator infection rates ($N_{\text{total \#egg sacs}}$ = 121). After collection, each egg sac was separated from the string, weighed, and stored in a temperature controlled room at 25 °C in falcon tubes with foam bungs. Egg sacs were misted twice weekly and monitored until spiderlings and/or *P. palanichamyi* emerged. Wasps were counted as they emerged, while photographs of spiderlings were taken for later counting due to high numbers emerging. We counted the spiderlings using the freely available software Dot Dot Goose (version 1.5.3) [47].

A further 96 egg sacs were collected from CM, AQ and SN in southern Spain in May 2022, and brought back to the lab. Here, they were kept at room temperature (~22 °C) and misted, as described above, and presence versus absence of emerging spiderlings and wasps was recorded over the following six weeks.

2.6. Statistics

All statistical analyses were conducted using R (version 4.1.1) [48]. The raw data is available in Table S1, SM1 Raw Data.

2.6.1. Spider Sizes

For each experimental colony, we calculated an average body mass, both pre- and post-experiments, as well as a colony-average body length and web size post-experiment. The distributions of all four variables were heavily left-skewed, so we used the non-parametric Spearman's rank correlation to test the correlations between the per-colony average values of pre- and post-weights, post-weight and post-web size, post-weight and post-body length, and post-body length and post-web size, all with N = 20. We used these correlations to justify interchangeably using body mass and web size as proxies for spider body size.

We further asked whether spiders had grown over the course of the experiment by testing the difference in average body mass pre- versus post-experiment with a Wilcoxon's test for matched pairs (N = 20). Due to the resulting significant growth of spiders over the experiment, which may have influenced their capture-abilities over time, we used the average between the pre- and post- average body masses as a response variable in the prey capture data analyses described below.

2.6.2. Control Prey Capture

In the control treatment where five fruit flies were introduced per colony five times, we asked whether larger spiders generally have higher prey capture success. We did this by testing the correlation between average spider body mass and the total number of flies caught per colony (up to a max. of 25 over the five feeding trials) using a Spearman's rank correlation test (N = 20).

2.6.3. Prey Capture Treatments

We investigated the effect of prey type and spider body size on prey capture success using a Generalised Linear Model (GLM) fitted with a binomial error structure and logit link function. We created a proportional response variable in the form of binding together two vectors, one that included the number of a prey type caught per colony over three trials (and so ranging from 0 to 3) and the second that included the number of prey items not caught (=3−the number caught).

As predictor variables, we included the per-colony average spider body mass, the treatment prey type, and their interaction term. The treatment prey type had four levels, as follows: soldier flies introduced singly, *T. absoluta* introduced singly, *T. absoluta* introduced singly together with a single fruit fly, and fruit flies introduced singly together with a single *T. absoluta* (N per treatment prey type = 20; total N in the model = 80).

We checked the distribution of prey caught to ensure a lack of zero inflation, and further tested to ensure a lack of overdispersion before proceeding to significance testing. Finally, we tested to ensure the full model was significant before proceeding to test the significance of the predictor variables. Significance of the interaction term and predictor variables were tested by comparing full models with reduced models.

We further ran the full model on a subset of the dataset that excluded the black soldier fly treatment. We did this as a post-hoc test to test for any differences in prey capture success between *T. absoluta* and easily caught fruit flies.

3. Results

3.1. Spider Sizes

All non-parametric correlations between proxies for spider body size were highly significant and positive: pre- and post-experiment average body mass (rho = 0.95, $p < 0.001$) post-experiment body mass and web size (rho = 0.93, $p < 0.001$), post-experiment body mass and body length (rho = 0.98, $p < 0.001$), and body length and web size (rho = 0.86, $p < 0.001$). These strong, positive correlations justify the interchangeable use of body mass, body length, and web size as equally valid proxies for spider size. Spiders were significantly heavier after the end of the experiment (Wilcoxon's test V = 2, $p < 0.001$); therefore, we used the average between the pre- and post-average body masses per colonies in the prey capture analyses, as described below.

3.2. Control Prey Capture

Larger spiders were able to capture significantly more prey in our control experiment, where multiple easily caught prey items (five wingless fruit flies) were introduced at a time (S = 58.1, $p < 0.001$, rho = 0.96, Figure 2a).

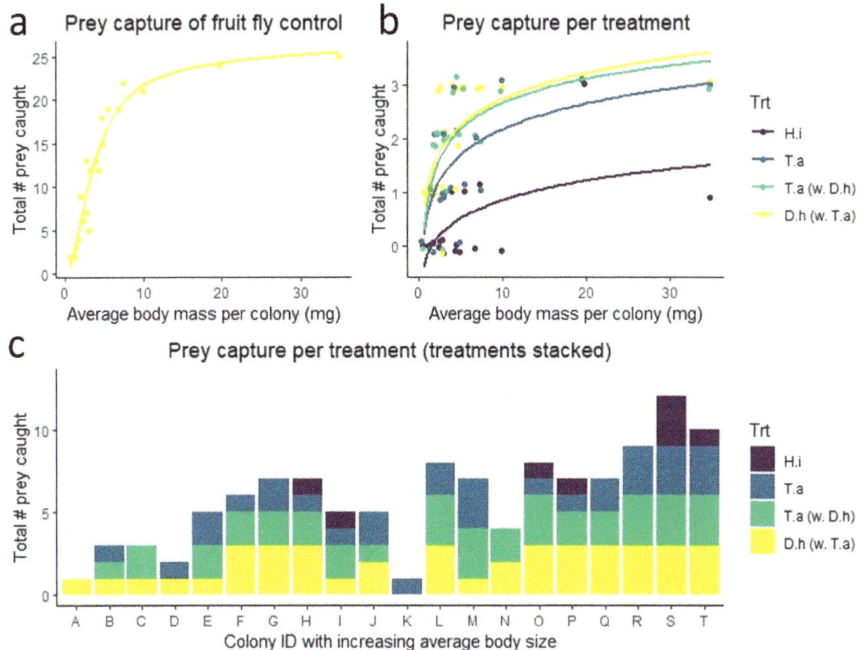

Figure 2. Prey capture results. (**a**) The total number of wingless fruit flies, *D. hydei*, caught after introducing five flies together, in each of five trials, in relation to the average spider body mass per colony (averaged over pre- and post-treatments weights); (**b,c**) The total number of prey items captured over three trials where prey was introduced singly: a black soldier fly alone (*H. illucens*) illustrated in purple; *T. absoluta* alone in blue; *T. absoluta* introduced together with a fruit fly in green; and a wingless fruit fly (*D. hydei*) introduced together with a *T. absoluta* in yellow.

3.3. Prey Capture Treatments

We found a significant interaction between spider size and prey type treatment (binomial GLM, p = 0.0032). This means that the general increase in prey capture success for larger spiders differed according to prey type (Figure 2b,c). Prey capture of *T. absoluta* was 100% for colonies of an average body mass of ~9 mg, body length of ~5 mm and web size of ~14 cm, and above, while spiders of a range of body sizes captured 100% of fruit flies, and spiders of most body sizes tended to be unsuccessful in capturing black soldier flies (Figure 2b,c).

In a posthoc test, where black soldier flies were excluded, the interaction between spider size and prey type treatment was not significant (binomial GLM, p = 0.25). Instead, spider size was a highly significant predictor of prey capture success (p < 0.001), whereas treatment was not significant (p = 0.058). Hence, whilst the capture success of *T. absoluta* was slightly lower than that of flightless fruit flies, this difference was not statistically significant, suggesting that spiders had no preference for either prey type.

3.4. Seasonality

Naturally occurring *C. citricola* webs in southern Spain fluctuated over the year according to the breeding season: as females grew and sexually matured in spring, webs grew larger and peaked in May and June, with most webs being 20–30 cm (Figure 3). After reproducing, most adult females died and the small webs of their offspring (≤10 cm) slowly started to dominate over the summer, although a few adult, breeding females were present year round. Offspring body sizes and, hence, web sizes began to increase over autumn and winter until the next main breeding season in spring.

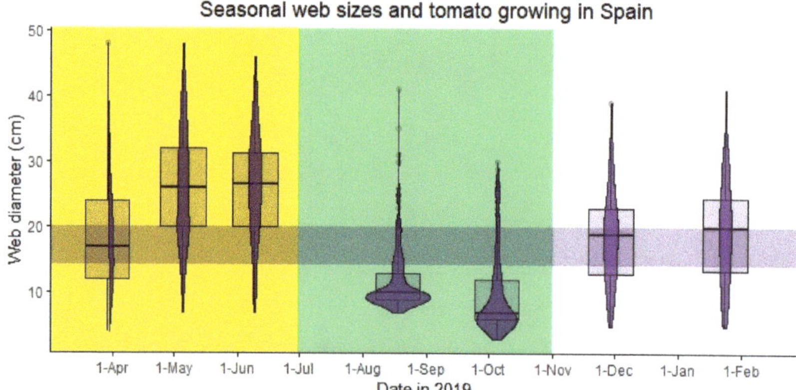

Figure 3. Web size seasonality and tomato growing in southern Spain. Violin plots overlaying box-and-whisker plots show the seasonal fluctuations in *C. citricola* web sizes in natural, field settings. The tomato planting and growing season is indicated in yellow (March–June) while the harvest season is indicated in green (July–Oct). A purple, horizontal band indicates the range of web sizes for which we found 100% *T. absoluta* prey capture success in our prey capture experiment in controlled lab settings.

In southern Spain, the tomato planting and growing season (March-June) coincides with the *C. citricola* main breeding season, when webs are at their large size, while webs found during the tomato harvest season, July–October, are at their smallest size (Figure 3) [49].

3.5. Wasp Infection

Out of 121 collected egg sacs in 2019, 12 had to be discarded because of labelling error. Out of the remaining 109 egg sacs, 73 produced live animals with an overall infection rate of 54.8%. Of the 73 egg sacs, 33 (45.2%) produced spiderlings only, with a median of 191 hatchlings (max. = 396, average = 181.4, st.dev = 87.5). Another 37 egg sacs (50.7%) were infected with wasps and produced zero spiderlings. From these egg sacs, a median of 18 wasps emerged (max. = 79, average = 26.5, st.dev. = 21.8). In only 3 egg sacs (4.1%), did some spiderlings survive a wasp infection, and these produced both spiderlings (between 46 and 104) and wasps (between 6 and 22).

From the 96 collected egg sacs in 2022, 73 egg sacs produced live animals, with an overall infection rate of 69.9%. Out of these, the proportion of egg sacs from which some spiderlings survived an infection was 42.5% (both spiderlings and wasps emerged from 31 egg sacs), while 22 egg sacs (30.1%) produced spiderlings only and 20 (27.4%) produced wasps only.

4. Discussion

This study set out to test the ability of the communal spider, *C. citricola*, to capture the tomato pest, *T. absoluta*, by providing laboratory-reared *C. citricola* colonies with different prey types. Small, experimental colonies of juvenile spiders were able to capture both the leafminers and flightless fruit flies, considered to be easily caught prey, with near-equal efficiency, and spiders showed no significant preference for either species. Larger black soldier flies, however, were rarely caught in these settings. This suggests that the spiders are as likely to capture and prey on leafminers as other small, easily caught insects, and, therefore, show promise as a potentially effective biological control agent of the moth. However, spider body size, which positively correlated with web size, was a strong predictor of prey capture success of all prey types tested. The capture success of the leafminer only reached 100% when juvenile spiders were about 5 mm in body length and had webs of roughly 14 cm in diameter and above, suggesting that hatchlings and very small spiderlings would be ineffective predators of adult, flying leafminers. Our experiments were conducted in the laboratory, and the spiders were exposed to prey items for three full days, which may also not fully represent prey capture dynamics in the field. Therefore, future studies should test prey capture efficacy in field settings, where prey may have a higher chance of avoiding capture webs.

Spiders in wild colonies of *C. citricola* in southern Spain produced the largest webs in May and June, which indicates that these months are the most opportune for the use of *C. citricola* as a biological control agent. This period neatly correlates with the beginning of the tomato growing season in Andalusia, southern Spain [49], where control of *T. absoluta* is crucial for commercial tomato farms. We found that web size is a function of both spider body mass and body length in *C. citricola*, meaning that prey capture potential of a spider colony could potentially be predicted by estimating average web sizes. Measuring individual spider web sizes in field settings is quick and easy [42], and would be an undisruptive method of gauging colony-level capture rate efficacy. This could be especially useful for predicting the pest-control capability of a developing colony of biological control spiders after introduction to an agricultural system. Nevertheless, it is important to note that the spiders used in this study were juveniles, and it is, therefore, unknown whether much larger, adult individuals would expend the effort to consume small *T. absoluta* individuals, especially when larger prey is likely to be accessible to them a natural setting [42]. While further research is needed to confirm that larger spiders (subadult and adult females) will also prey on *T. absoluta*, we know from previous studies that larger spiders often catch relatively small prey, and thus, are likely to prey on the relatively small moths. In Grinsted et al. (2019), larger spiders, including adults, with webs between 20 and 37 cm in diameter preyed mainly on insects smaller than the average *T. absoluta* body length of 6 mm [50]. Indeed, median prey length for these larger females in natural field settings was 3 mm (prey body length ranged from 1–17 mm, with 75% of prey <6 mm), calculated from the raw data deposited by Grinsted et al. (2019) [42]. These results also suggest that spiders in Spain are unlikely to be effective as pest control agents during the harvest season in southern Spain [49], as spider webs are mostly too small during July–October. This further suggests that seasonal fluctuations in web sizes in a given geographical region must be taken into consideration prior to *C. citricola* application [44]. Despite the spiders' efficacy at catching the tomato pest, it is important to consider that they are generalist predators, and are, therefore, capable of removing beneficial pollinators [7,18,44], which are crucial to the fertilisation of the tomato crop. It is, therefore, integral that the effects of biological control colonies of spiders on pollinator populations are considered in future studies.

When assessing the efficacy of a novel biological control agent, it is imperative that community ecology is considered, as interactions with other species in the community may hamper pest control abilities [29,33,38]. One possible disruptor of the effectiveness of *C. citricola* as a biological control agent is egg predation by the wasp, *P. palanichamyi*, a species that oviposits into the egg sacs of *C. citricola*, and emerging larvae consume the developing spider eggs [25]. We found an infection rate of >50% of egg sacs at our field sites near Cádiz, while Chuang et al. (2019) [25] found that about 42% of egg sac strings were infected over a larger area of southern Spain, from Cádiz to Valencia. We found large variations in spiderling survival after wasp infections, but overall, ~30–50% of sampled egg sacs produced wasps only, with zero surviving spiderlings. Hence, wasp infections may cause severe predation pressure and possibly shape extinction patterns in *C. citricola*, at least in Spain, as suggested by Chuang et al. (2019) [25]. Furthermore, by introducing high numbers of *C. citricola* to an area as a biological control agent, more egg sacs will subsequently be provided as prey for *P. palanichamyi*, resulting in population increase of the wasp, and possible local community ecology alteration [51,52]. Additionally, the implementation of additional *C. citricola* colonies into ecosystems is also likely to facilitate population expansion of both the colony-associate *H. pluchei* and kleptoparasitic *Argyrodes* spp., which could also potentially detrimentally affect *C. citricola* populations.

It is, therefore, important to address how community ecology may impact *C. citricola* population dynamics and the resulting pest control efficacy [52]. Thus, we propose that species communities within spider colonies, particularly focussing on *H. pluchei* and *A. argyrodes*, as well as wasp infection rates in nearby rural populations, should be closely monitored during implementation of the biological control agent. Furthermore, in the event of wasp populations expanding and causing potential harm to both natural and biological control spider populations [52], a plan to control wasp infections must be devised. A possible future avenue for research is therefore testing the rate of increase in *P. palanichamyi* populations over multiple generations, in the presence of increasing egg sac numbers.

The proposed efficacy of *C. citricola* as a biological control agent is based on two useful facets of the species. Firstly, the evolution of group living and high conspecific tolerance confers reduced aggression toward neighbouring spiders, and therefore, fewer incidences of intraspecific attack and cannibalism as compared to solitary spiders [32,37,42]. This is likely to result in high predator density when used as a biological control agent and the ability to intercept large numbers of pest arthropods with their interconnected capture webs [37,42]. Furthermore, few studies have tested the efficacy of group-living spiders as biological control agents, many focussing on comparing web-building and non-web-building spiders, despite the aforementioned advantages of the use of communal species [44,52,53]. Secondly, the global ubiquity of *C. citricola* may result in its potential use in multiple locations worldwide, including LMICs such as those in Mediterranean Europe, Africa, Asia, and the Middle East, whose environmental health and economic stability could greatly benefit from this sustainable agricultural approach [25,27]. However, incidences of the spider becoming an agricultural pest have occurred in countries including the Dominican Republic and Columbia, where it was introduced in the last 25 years [54]. Here, colonies can comprise a great number of individuals, and excessive colony expansion results in increased capture web construction, which has been reported to asphyxiate crop plants, resulting in reduced crop yield and potentially causing economic deficit [25,54,55]. It is, therefore, important to ensure that appropriate substrate is provided for spiders to build their colonies on, and that spiders are discouraged from building webs directly on the crops, before commercial use can commence [7]. Such substrates could be natural supports, such as the *Opuntia* spp. cactus, which is both a favoured host of wild *C. citricola* colonies, and a commonly grown border plant around agricultural fields in southern Spain [43]. Inorganic frames could also be used; however, these may be less effective than preferred host plants, such as *Opuntia* cacti, as modified cactus stems provide wind protection that increases prey vibration sensitivity [43] and are, therefore, likely to benefit prey capture rate in a way that wire supports could not. In cases where introduction of *Opuntia* field borders is not possible, suitable inorganic substrates must be developed to ensure that *C. citricola* pest control colonies can reach their highest potential prey capture efficiency.

In this study, we noted multiple instances of *T. absoluta* sitting uncaptured in *C. citricola* webs in all colonies during interim checks during the three-day capture period, either suggesting that the moth is not always detectable by the spiders, or that the moths can avoid becoming caught in capture webs. It may be possible that the low body mass of *T. absoluta* results in the production of few vibratory signals, causing inconsistency in prey capture capability. In addition, moth scales, which are likely to be lost as they brush against the non-adhesive silk strands of *C. citricola* webbing, may allow them to avoid becoming trapped [56]. Further research is needed to ascertain whether this phenomenon could affect the biological control capability of the spider.

One limitation of these spiders as potential biological control agents is that their capture webs are specialised to capture arthropods that fly, jump, or fall into them, and are, therefore, unable to control pests at larval instars. The inclusion of the spider into an Integrated Pest Management (IPM) system, such as pairing it with entomopathogenic nematodes, which are already marketed as biological control agents of phytophagous larvae [17,24,57], could improve its efficacy as a biological control agent. It has been suggested that *Steinernema feltiae* nematodes may have the propensity to facilitate up to

68% mortality in larval *T. absoluta* [57]. This combination could potentially control the leafminer at both larval and adult instars, negating the shortcomings of both constituents of the management system. This approach may also reduce pesticide reliance in agricultural tomato crops, resulting in reduced pollutants in soils, waterways, and food chains [15].

5. Conclusions

In this study, we found that facultatively group-living *C. citricola* spiders caught leafminers and flightless fruit flies at the same rate in lab-based trials, and that larger capture web production coincides with the tomato planting and growing season in southern Spain, suggesting that this communal spider could be a potentially successful candidate for use as a biological control agent of *T. absoluta*. These findings open doors for the use of group-living arachnids to control agricultural pests, reducing commercial pesticide dependence, and having significant beneficial outcomes for environmental and economic stability, particularly in LMICs [14]. Furthermore, many LMICs exist within the overlapping geographic ranges of both *C. citricola* and *T. absoluta* [10,19,28], meaning that the introduction of pest control spiders in these regions will be unlikely to significantly damage native biodiversity. Although these results are promising, downsides to the use of spiders as pest control agents still remain; the two main issues raised being: (a) that spiders are generalist predators and are likely to catch integral tomato-pollinating arthropods [7,18,44]; and (b) that increasing spider populations will also alter community ecology, and may result in increasing predator and kleptoparasite densities [25,29,33,38]. Therefore, it is crucial that community ecology is monitored after the introduction of biological control spiders in order to preserve the health of the ecosystem and to ensure that maximum biological control efficacy is maintained. Future studies are now needed to test the efficacy of *C. citricola* for pest control in field setting, and to test the potential of other promising group-living spider species to provide pest control [44,58,59].

Supplementary Materials: The following supporting information can be downloaded at: https://www.mdpi.com/article/10.3390/insects14010034/s1. Prey capture experiments, lab spider sizes, field web sizes, 2019 wasp infections, 2022 wasp infections, prey body masses.

Author Contributions: Conceptualization, L.G. and T.A.R.-M.; methodology, T.A.R.-M., L.G. and E.K.D.; formal analysis, L.G.; investigation, T.A.R.-M., E.K.D. and M.A.M.; data curation, T.A.R.-M., E.K.D. and L.G.; writing—original draft preparation, T.A.R.-M.; writing—review and editing, L.G., E.K.D. and M.A.M.; visualization, L.G.; supervision, L.G.; funding acquisition, L.G. All authors have read and agreed to the published version of the manuscript.

Funding: This research was funded by the Royal Society, grant number RSG\R1\180271. M.A.M. was supported by a Zuckerman STEM Postdoctoral fellowship. A permit was obtained from Junta de Andalucía for the collection and export of spider and wasp specimens from Spain, with reference number GB-58/19. A permit for the export of *T. absoluta* from the Republic of Ghana was obtained from the Ministry of Food and Agriculture, Plant Protection & Regulatory Services Directorate, PC No: 20/0003955. UK imports were declared to APHA & DEFRA with IPAFFS Reference numbers CHEDA.GB.2021.1336106; IMP.GB.2021.1202482; IMP.GB.2022.1988376.

Institutional Review Board Statement: Ethical approval was granted by Animal Welfare and Ethical Review Body (AWERB), University of Portsmouth on 6 May 2022. Confirmation code: 521A.

Data Availability Statement: The data presented in this study are available in the Supplementary Materials entitled SM1 Raw Data.

Acknowledgments: We would like to thank Manuel Jimenez-Tenorio and Juan Carlos Garcia Galindo for collecting the *C. citricola* egg sacs used in the laboratory-based aspect of this study, as well as Ian Baxter of BioBest, Chris Rose of R&L Holt's, Lakpo Agboyi of CABI, and Mark Wilde, for help providing the *T. absoluta* used in the study. Thanks to Christine Hughes for helping care for the spiders in the lab, and to Lauren Brown, Joshua Goacher and Madeline Walker for lab assistance.

Conflicts of Interest: The authors declare no conflict of interest. The funders had no role in the design of the study; in the collection, analyses, or interpretation of data; in the writing of the manuscript; or in the decision to publish the results.

References

1. Bebber, D.P.; Ramotowski, M.A.T.; Gurr, S.J. Crop pests and pathogens move polewards in a warming world. *Nat. Clim. Chang.* **2013**, *3*, 985–988. [CrossRef]
2. Campos, M.R.; Biondi, A.; Adiga, A.; Guedes, R.N.C.; Desneux, N. From the Western Palaearctic region to beyond: *Tuta absoluta* 10 years after invading Europe. *J. Pest Sci.* **2017**, *90*, 787–796. [CrossRef]
3. Arnó, J.; Molina, P.; Aparicio, Y.; Denis, C.; Gabarra, R.; Riudavets, J. Natural enemies associated with *Tuta absoluta* and functional biodiversity in vegetable crops. *BioControl* **2021**, *66*, 613–623. [CrossRef]
4. Tarusikirwa, V.L.; Machekano, H.; Mutamiswa, R.; Chidawanyika, F.; Nyamukondiwa, C. *Tuta absoluta* (Meyrick) (Lepidoptera: Gelechiidae) on the "offensive" in Africa: Prospects for integrated pest management initiatives. *Insects* **2020**, *11*, 764. [CrossRef]
5. Gebremariam, G. *Tuta absoluta*: A global looming challenge in tomato production. *J. Biol. Agric. Healthc.* **2015**, *5*, 57–63.
6. Tropea Garzia, G.; Siscaro, G.; Biondi, A.; Zappalà, L. *Tuta absoluta*, a South American pest of tomato now in the EPPO region: Biology, distribution and damage. *EPPO Bull.* **2012**, *42*, 205–210. [CrossRef]
7. Michalko, R.; Pekár, S.; Entling, M.H. An updated perspective on spiders as generalist predators in biological control. *Oecologia* **2019**, *189*, 21–36. [CrossRef]
8. Desneux, N.; Wajnberg, E.; Wyckhuys, K.A.G.; Burgio, G.; Arpaia, S.; Narváez-Vasquez, C.A.; González-Cabrera, J.; Ruescas, D.C.; Tabone, E.; Frandon, J.; et al. Biological invasion of European tomato crops by *Tuta absoluta*: Ecology, geographic expansion and prospects for biological control. *J. Pest Sci.* **2010**, *83*, 197–215. [CrossRef]
9. Chakraborty, S.; Newton, A.C. Climate change, plant diseases and food security: An overview. *Plant Pathol.* **2011**, *60*, 2–14. [CrossRef]
10. Ferracini, C.; Bueno, V.H.P.; Dindo, M.L.; Ingegno, B.L.; Luna, M.G.; Salas Gervassio, N.G.; Sánchez, N.E.; Siscaro, G.; van Lenteren, J.C.; Zappalà, L.; et al. Natural enemies of *Tuta absoluta* in the Mediterranean basin, Europe and South America. *Biocontrol Sci. Technol.* **2019**, *29*, 578–609. [CrossRef]
11. Van Den Bosch, R. Biological control of insects. *Annu. Rev. Ecol. Syst.* **1971**, *2*, 45–66. [CrossRef]
12. Buragohain, P.; Saikia, D.K.; Sotelo-Cardona, P.; Srinivasan, R. Evaluation of bio-pesticides against the south american tomato leaf miner, *Tuta absoluta* (Meyrick) (Lepidoptera: Gelechiidae) in India. *Horticulturae* **2021**, *7*, 325. [CrossRef]
13. Guedes, R.N.C.; Picanço, M.C. The tomato borer *Tuta absoluta* in South America: Pest status, management and insecticide resistance. *EPPO Bull.* **2012**, *42*, 211–216. [CrossRef]
14. Ecobichon, D.J. Pesticide use in developing countries. *Toxicology* **2001**, *160*, 27–33. [CrossRef] [PubMed]
15. Terziev, V.; Petkova-Georgieva, S. The pesticides toxic impact on the human health condition and the ecosystem. *SSRN Electron. J.* **2019**, *5*, 1314–1320. [CrossRef]
16. Lushchak, V.I.; Matviishyn, T.M.; Husak, V.V.; Storey, J.M.; Storey, K.B. Pesticide toxicity: A mechanistic approach. *EXCLI J.* **2018**, *17*, 1101–1136. [CrossRef]
17. Dlamini, B.E.; Dlamini, N.; Masarirambi, M.T.; Nxumalo Kwanele, A. Control of the tomato leaf miner, *Tuta absoluta* (Meyrick) (Lepidoptera: Gelechiidae) larvae in laboratory using entomopathogenic nematodes from subtropical environment. *J. Nematol.* **2020**, *52*, 1–8. [CrossRef]
18. Yanira Olivera, S.; Arroyo, N.; Asencio, C. *Cyrtophora citricola* as a potential component of an integrated pest management program for citrus in Puerto Rico. *FASEB J.* **2013**, *27* (Suppl. 1). [CrossRef]
19. Desneux, N.; Han, P.; Mansour, R.; Arnó, J.; Brévault, T.; Campos, M.R. Integrated pest management of *Tuta absoluta*: Practical implementations across different world regions. *J. Pest Sci.* **2022**, *95*, 17–39. [CrossRef]
20. Gonzalez-Cabrera, J.; Molla, O.; Monton, H.; Urbaneja, A. Efficacy of *Bacillus thuringiensis* (Berliner) in controlling the tomato borer, *Tuta absoluta* (Meyrick) (Lepidoptera: Gelechiidae). *BioControl* **2011**, *56*, 71–80. [CrossRef]
21. Mascarin, G.M.; Alves, S.B.; Rampelotti-Ferreira, F.T.; Urbano, M.R.; Demétrio, C.G.B.; Delalibera, I. Potential of a granulovirus isolate to control *Phthorimaea operculella* (Lepidoptera: Gelechiidae). *BioControl* **2010**, *55*, 657–671. [CrossRef]
22. Aigbedion-Atalor, P.O.; Mohamed, S.A.; Hill, M.P.; Zalucki, M.P.; Azrag, A.G.A.; Srinivasan, R.; Ekesi, S. Host stage preference and performance of *Dolichogenidea gelechiidivoris* (Hymenoptera: Braconidae), a candidate for classical biological control of *Tuta absoluta* in Africa. *Biol. Control* **2020**, *144*, 104215. [CrossRef]
23. Contreras, J.; Mendoza, J.E.; Martínez-Aguirre, M.R.; García-Vidal, L.; Izquierdo, J.; Bielza, P. Efficacy of enthomopathogenic fungus *Metarhizium anisopliae* against *Tuta absoluta* (Lepidoptera: Gelechiidae). *J. Econ. Entomol.* **2014**, *107*, 121–124. [CrossRef] [PubMed]
24. Kamali, S.; Karimi, J.; Koppenhöfer, A.M. New insight into the management of the tomato leaf miner, *Tuta absoluta* (Lepidoptera: Gelechiidae) with entomopathogenic nematodes. *J. Econ. Entomol.* **2018**, *111*, 112–119. [CrossRef]
25. Chuang, A.; Gates, M.W.; Grinsted, L.; Askew, R.; Leppanen, C. Two hymenopteran egg sac associates of the tent-web orbweaving spider, *Cyrtophora citricola* (Forskål, 1775) (Araneae, Araneidae). *ZooKeys* **2019**, *874*, 1–18. [CrossRef]

26. Lubin, Y.; Bilde, T. The evolution of sociality in spiders. *Adv. Study Behav.* **2007**, *37*, 83–145. [CrossRef]
27. World Spider Catalog. Available online: https://wsc.nmbe.ch/species/3773 (accessed on 17 October 2022).
28. Urbaneja, A.; González-Cabrera, J.; Arnó, J.; Gabarra, R. Prospects for the biological control of *Tuta absoluta* in tomatoes of the Mediterranean basin. *Pest Manag. Sci.* **2012**, *68*, 1215–1222. [CrossRef]
29. Rypstra, A.L. Foraging flocks of spiders. *Behav. Ecol. Sociobiol.* **1979**, *5*, 291–300. [CrossRef]
30. Yip, E.C.; Rao, D.; Smith, D.R.; Lubin, Y. Interacting maternal and spatial cues influence natal—Dispersal out of social groups. *Oikos* **2019**, *128*, 1793–1804. [CrossRef]
31. Yip, E.C.; Levy, T.; Lubin, Y. Bad neighbors: Hunger and dominance drive spacing and position in an orb-weaving spider colony. *Behav. Ecol. Sociobiol.* **2017**, *71*, 128. [CrossRef]
32. Johannesen, J.; Wennmann, J.T.; Lubin, Y. Dispersal behaviour and colony structure in a colonial spider. *Behav. Ecol. Sociobiol.* **2012**, *66*, 1387–1398. [CrossRef]
33. Leborgne, R.; Cantarella, T.; Pasquet, A. Colonial life versus solitary life in *Cyrtophora citricola* (Araneae, Araneidae). *Insectes Soc.* **1998**, *45*, 125–134. [CrossRef]
34. Yip, E.C.; Smith, D.R.; Lubin, Y. Long-term colony dynamics and fitness in a colonial tent-web spider *Cyrtophora citricola*. *Front. Ecol. Evol.* **2021**, *9*, 725647. [CrossRef]
35. Uetz, G.W. The "ricochet effect" and prey capture in colonial spiders. *Oecologia* **1989**, *81*, 154–159. [CrossRef] [PubMed]
36. Su, I.; Buehler, M.J. Mesomechanics of a three-dimensional spider web. *J. Mech. Phys. Solids* **2020**, *144*, 104096. [CrossRef]
37. Grinsted, L.; Lubin, Y. Spiders: Evolution of group living and social behavior. In *Encyclopedia of Animal Behavior*; Choe, J.C., Ed.; Academic Press: London, UK, 2019; pp. 632–640. [CrossRef]
38. Sunderland, K. Mechanisms underlying the effects of spiders on pest populations. *J. Arachnol.* **1999**, *27*, 308–316.
39. Wise, D.H. Cannibalism, food limitation, intraspecific competition, and the regulation of spider populations. *Annu. Rev. Entomol.* **2006**, *51*, 441–465. [CrossRef]
40. Nyffeler, M. Prey selection of spiders in the field. *Am. Arachnol. Soc.* **1999**, *27*, 317–324.
41. Marc, P.; Canard, A.; Ysnel, F. Spiders (Araneae) useful for pest limitation and bioindication. *Agric. Ecosyst. Environ.* **1999**, *74*, 229–273. [CrossRef]
42. Grinsted, L.; Deutsch, E.K.; Jimenez-Tenorio, M.; Lubin, Y. Evolutionary drivers of group foraging: A new framework for investigating variance in food intake and reproduction. *Evolution* **2019**, *73*, 2106–2121.
43. Chuang, A.; Leppanen, C. Location and web substrate records of *Cyrtophora citricola* (Araneae: Araneidae) in southern Spain, including Tenerife. *Rev. Ibérica Aracnol.* **2018**, *33*, 89–100.
44. Riechert, S.E.; Lockley, T. Spiders as biological control agents. *Annu. Rev. Entomol.* **1984**, *29*, 299–320. [CrossRef]
45. Yip, E.C.; Berner-Aharon, N.; Smith, D.R.; Lubin, Y. Coy males and seductive females in the sexually cannibalistic colonial spider, *Cyrtophora citricola*. *PLoS ONE* **2016**, *11*, e0155433. [CrossRef] [PubMed]
46. Grinsted, L.; Schou, M.F.; Settepani, V.; Holm, C.; Bird, T.L.; Bilde, T. Prey to predator body size ratio in the evolution of cooperative hunting—A social spider test case. *Dev. Genes Evol.* **2020**, *230*, 173–184. [CrossRef] [PubMed]
47. Ersts, P.J. *DotDotGoose*, Version 1.5.3; American Museum of Natural History: New York, NY, USA, 2022. Available online: https://biodiversityinformatics.amnh.org/open_source/dotdotgoose (accessed on 17 October 2022).
48. RStudio Team. *RStudio: Integrated Development Environment for R*; PBC: Boston, MA, USA, 2022.
49. Branthôme, F.-X. Spain: The Other European Giant. Available online: https://www.tomatonews.com/en/spain-the-other-european-giant_2_264.html (accessed on 1 February 2022).
50. Visser, D.; Uys, V.M.; Nieuwenhuis, R.J.; Pieterse, W. First records of the tomato leaf miner *Tuta absoluta* (Meyrick, 1917) (lepidoptera: Gelechiidae) in South Africa. *BioInvasions Rec.* **2017**, *6*, 301–305. [CrossRef]
51. Tellier, A.; Brown, J.K.; Boots, M.; John, S. Theory of host-parasite coevolution: From ecology to genomics. *eLS* **2021**, *2*, 1–10. [CrossRef]
52. Riechert, S.E. The hows and whys of successful pest suppression by spiders: Insights from case studies. *J. Arachnol.* **1999**, *27*, 387–396.
53. Michalko, R.; Pekár, S.; Dul'a, M.; Entling, M.H. Global patterns in the biocontrol efficacy of spiders: A meta-analysis. *Glob. Ecol. Biogeogr.* **2019**, *28*, 1366–1378. [CrossRef]
54. Edwards, G.B. *Cyrtophora citricola* (Araneae: Araneidae): A colonial tentweb orbweaver established in Florida. *Entomol. Circ. Fla. Dept. Agric. Consum. Serv.* **2006**, *411*, 1–5.
55. Stange, L.A. A colonial tentweb orbweaver *Cyrtophora citricola*. *Edis* **2013**, *2013*, 1–6. [CrossRef]
56. Diaz, C.; Maksuta, D.; Amarpuri, G.; Tanikawa, A.; Miyashita, T.; Dhinojwala, A.; Blackledge, T.A. The moth specialist spider *Cyrtarachne akirai* uses prey scales to increase adhesion. *J. R. Soc. Interface* **2020**, *17*, 20190792. [CrossRef]
57. Ben Husin, T.O.; Port, G.R. Efficacy of entomopathogenic nematodes against *Tuta absoluta*. *Biol. Control* **2021**, *160*, 104699. [CrossRef]

58. Mishra, A.; Rastogi, N. Unraveling the roles of solitary and social web-making spiders in perennial ecosystems: Influence on pests and beneficials. *Proc. Natl. Acad. Sci. India Sect. B Biol. Sci.* **2020**, *90*, 567–576. [CrossRef]
59. Grinsted, L.; Agnarsson, I.; Bilde, T. Subsocial behaviour and brood adoption in mixed-species colonies of two theridiid spiders. *Naturwissenschaften* **2012**, *99*, 1021–1030. [CrossRef] [PubMed]

Article

Effects of Wolf Spiders' Captive Environment on Their Locomotor and Exploratory Behaviours

Marie Trabalon

EthoS-UMR 6552, CNRS, Université de Rennes 1, F-35000 Rennes, France; marie.trabalon@univ-rennes1.fr; Tel.: +33-223-237-101

Simple Summary: Since the 1960s, abuses of domestic and of wild animals that have been tamed or are held in captivity have been legally prohibited and laws ensure their well-being. Many scientific investigations carried out in this context recommend ways to adapt farming and thus to avoid physical and/or psychological suffering. Evaluations of animals' welfare in captivity entail the need to understand in detail the fundamental behaviours of the focus species and to know the degree of their variation to be able to establish objective bases that can ensure breeding conditions that respect the animals' welfare. Current laws do not apply to invertebrate animals (such as insects or spiders) and consideration of the welfare of these animals in captivity is neglected. Here, I compared the behaviour of wild adult spiders just after collection and that of adult spiders hatched and bred in the laboratory. My results show that captivity induced rapid changes of wild spiders' behaviour once in captivity. Therefore, it is important to establish the best breeding conditions for the needs of both invertebrate and vertebrate animals in order to promote their well-being.

Abstract: Here I detail the effects of the abiotic/captive environment of an adult wandering spider, *Pardosa saltans* (*Lycosidae*) on its behaviour. These studies focused on spiders collected as adults in their natural environment and spiders developed in the laboratory under controlled conditions. Wild-caught spiders were tested either immediately after capture or after being housed for 15 days post-collection. Laboratory reared spiders were kept in different environments: small or large space combined with the presence or absence of litter. Two tests evaluated by sex show the influence of these rearing conditions: an open-field test and a radial-arm maze test. The results show that wild caught spiders of both sexes tested immediately after capture weighed significantly less and were significantly more active than spiders housed in the laboratory for 15 days and spiders reared in the laboratory. Laboratory conditions induced a positive impact on body mass and negative impact on behaviour activities. The locomotor and exploratory activities of spiders of both sexes kept in container without substrate showed lower. My results suggest that the physical enrichment of the environment can reduce these negative effects for females, but not for males that seem to be more affected by being reared under controlled conditions.

Keywords: activity; experience; natural and artificial environments; male and female adult spiders

---

Citation: Trabalon, M. Effects of Wolf Spiders' Captive Environment on Their Locomotor and Exploratory Behaviours. *Insects* **2022**, *13*, 135. https://doi.org/10.3390/insects13020135

Academic Editors: Thomas Hesselberg and Dumas Gálvez

Received: 15 December 2021
Accepted: 25 January 2022
Published: 27 January 2022

Publisher's Note: MDPI stays neutral with regard to jurisdictional claims in published maps and institutional affiliations.

1. Introduction

For a few decades now, consideration of animal welfare has been growing in scientific circles, the food industry, legislation, and even at private and individual levels. In this context have emerged many studies focusing on the impact of captivity and breeding conditions on the well-being and, more generally, on the behavioural characteristics of many vertebrates [1,2]. Thus, it has been general knowledge for some time now that the enrichment of an individual's environment is an important factor that must be taken into consideration to ensure that an individual's coping skills and physical health are good and that it develops a rich behavioural repertoire [3,4]. These reports show that the well-being of domestic animals as well as that of animals in captivity can be improved by enriching

---

and complexifying their captive environment and by providing sufficient space. Levels of locomotor and exploratory activities are good indicators of an animal's well-being. Thus, a way to estimate the level of well-being of animals in captivity is to compare their levels of locomotor and exploratory activities to those of animals in their natural environment. Studies of this type but focusing on invertebrates are rare. However, we know that, as for vertebrates, environmental constraints experienced during ontogeny can have important effects on the morphological and physiological development of an invertebrate, as well as on the establishment of its behavioural repertoire, as for example its locomotor and exploratory behaviours.

Locomotor and exploratory activities reveal the tendency of all animals to move about and inspect their environment, even when neither hunger, nor thirst, nor sexual appetite compels them to do so. These behaviours allow an individual to gather a certain amount of information concerning their environment and are adapted to its living environment. In the event of a change in its environment during its lifetime an individual has to adapt its behaviour to the novel characteristics of its environment, an acclimatization that may involve a change in its behaviour. According to the literature, locomotor and exploratory activities are subject to environmental constraints (abiotic and biotic), genetic constraints (sexual dimorphism) and maturation constraints of an organism (physiological state). For example, a decrease in ambient temperature results in a decrease of the locomotor activities of most arthropods [5] and the presence of conspecifics in the environment induces an increase in their exploratory behaviour [6,7].

Sexual maturity causes a modification of the exploratory and locomotor activities of the majority of arthropod species and exploratory behaviours can then differ between adult males and females (spiders: [8]; myriapods and ground beetles: [9,10]; woodlouse: [11]). Generally, adult male arthropods move about more and this tendency can be explained by the fact that males, in addition to foraging, are looking for a sexual partner [12]. Conversely, females move about less, remaining almost in the same place to allow males to find them more easily [13].

Over the past years, a few authors have been interested in the impact of the rearing environment on different species of spiders [14]. Thus, some studies compare the locomotor and exploratory activities of individuals captured in their natural environment to that of individuals reared in an artificial environment. For example, *Phidippus audax* (*Salticidae*) spiders captured in the wild approach artificial prey faster and from a more distant position and are less static than individuals that had hatched and been reared in the laboratory [15–17]. Folz [18] showed that *Hogna helluo* (*Lycosidae*) spiders reared in smaller terraria were faster to search for and capture prey. Other observations showed that individuals that had developed in a physical environment enriched were more active than individuals maintained in a non-enriched environment [6,15,19,20].

My study aims to highlight differences between wandering spiders that developed in the laboratory and spiders that had developed in their natural environment. I hypothesised that the locomotor and exploratory behaviours of spiders reared in the laboratory would differ from those of spiders that had developed in their natural environment because of restricted living conditions in the laboratory. In addition, I evaluated the effects of different types of rearing conditions on the behaviour of female and male adults to evidence factors that could improve their living conditions. In this context, I questioned whether the size of their terraria would modify the spiders' behaviour in relation to the volume in which they had developed. I assumed that spiders reared in large terraria would be more active than those reared in small terraria. Second, I questioned whether the presence of litter in a terrarium would induce any behavioural changes. I assumed that a complex environment such as undergrowth litter would induce individuals to move about and explore their environment more. Third, I evaluated how long in captivity after being collected the behaviour of wild caught adult spiders would be modified. The subjects of most laboratory studies on reproduction behaviours (sexual and maternal) are individuals that have been kept in the laboratory for a long time or under maintenance conditions that affect individuals' behaviour. I assumed that 15 days of captivity would not lead to any behavioural modification. My model spider species is the free-moving wolf spider, *Pardosa saltans* (Lycosidae).

2. Materials and Methods

2.1. Ethics Statement

By using an invertebrate species and caring for it while using the accepted ethical standards in the laboratory, my research conforms to the legal requirements and guidelines established for the treatment of animals in research. The species used for these experiments is neither endangered nor protected.

2.2. Spider Collection and Rearing

Members of the species *Pardosa saltans* can be found in forests, woodlands and thickets, and sometimes near grasslands and hedges. Their distribution appears to be largely restricted to old and ancient woodland sites, where they can become numerous, running over the ground in open clearings as well as amongst litter in the shade of a wood. The breeding season of *P. saltans* extends from March to September in France [21]. During the day, *P. saltans* spiders move ceaselessly over the substrate in search of food, a sexual partner or a sunny area to warm up. These spiders hide under the litter at the end of the day and until the morning to protect themselves from high temperature variations and predators. It is for these various reasons that *Pardosa saltans* is an ideal model to study locomotor and exploratory behaviour.

Two experimental groups were used: spiders that had emerged and developed in their natural environment (natural spider group) and spiders that had emerged and been reared in the laboratory (laboratory spider group).

Natural spider group: The subjects were female and male adults captured in a private forest near Guichen (France; 47°58′03″ N, 1°47′43″ W) in April-May 2021. Spiders were randomly assigned to two experimental groups (Figure 1). Adults of first group were weighed (Sartorius electronic balance, ±0.01 g; Palaiseau, France) then tested between 2 and 3 h after collection and returned to their natural environment at the end of the behavioural tests (NE group). Adults of second group were weighed just after collection and then kept individually during 15 days in circular terrarium (15 cm in diameter, 5 cm high) with water and without soil, under the same temperature as outdoors during these days (22 ± 2 °C during day and 16 ± 3 °C during night; and hygrometry (37 ± 5% relative humidity) and natural photoperiodic cycle (NE-15 group). These NE-15 spiders were fed every five days ad libitum with juvenile crickets (*Acheta domestica*) alternating with adult flies (*Delia radicum*). After 15 days in captivity, spiders were weighed, behavioural tested and then returned to their natural environment.

Laboratory spiders groups: Juveniles spiders (*n* = 120) emerged from six cocoons in the laboratory during September 2020 and were reared in the laboratory until they were adult (April–May 2021). The juvenile spiders (7–8 days after emergence), after they had caught their first prey, where kept individually either in large (L × l: 17 × 9 cm, 8 cm in high) or small transparent terraria (L × l: 9 × 6 cm, 5 cm in high). Both large (L) and small (S) terraria were divided into two subgroups: either with 1 cm litter (mixture of earth and leaves from their native forest) on the base, or without any matter on the base (Figure 1). These spiders were kept for seven months under natural temperature (20 ± 4 °C during day and 10 ± 3 °C during night), hygrometry (37 ± 5% relative humidity) and natural photoperiodic cycle. They were fed every six days ad libitum, with juvenile crickets (*Acheta domestica* and *Nemobius sylvestris*) alternating with adult flies (*Delia radicum*). After seven months of development, all spiders were checked every day to record their adult moult and tested between 2 and 3 weeks after their adult moult.

Figure 1. Rearing conditions of adult spiders: developed in natural environment (NE group); wild caught spiders housed in the laboratory for 15 days (NE-15 group); juveniles spiders developed in the laboratory in large terraria without matter on the base until adult (L group); developed in large terraria with matter on the base until adult (LM group); developed in small terraria without matter on the base until adult (S group); developed in small terraria with matter on the base until adult (SM group).

2.3. Locomotor and Behaviour Activities of the Natural Spider Group in the Laboratory

The activities of the natural spiders ($n = 25$/sex) were video recorded in their terrarium the day they were collected just after they had been placed in their terrarium (day 0 in captivity) then 7 (7 days in captivity) and 14 days after collection (14 days in captivity), by a camera (video tracking). All observations were recorded between 14.00 and 17.00 h in natural light with the SMART-MA software program (Smart Panlab, Bioseb, France). The software program calculated directly locomotor activity (time of mobility, trajectories) expressed in surface explored (cm/min) (Table 1). Movements without changing place were recorded directly by the experimenter using a keyboard event recorder integrated in SMART-Ma system (key were designated as behaviour events). Six spontaneous behavioural events without moving were observed and recorded (Table 1). The program allows recording duration of events versus frequency of events. Timing of events is accurate to 0.1 s. During captivity in a terrarium, "abdominal" vibrations were manifested only by females and "drumming" only by males.

Table 1. Parameters measured for evaluation of locomotor, exploratory and spontaneous activities of a spider in its terrarium and during behavioural tests in a novel environment (open-field arena and radial-arm maze).

Behavioural Events	Parameters Measured
Locomotor activity (Surface explored in cm/min)	Time of mobility Trajectory
Spontaneous activity without displacement (Frequency of behaviour)	"Inactivity" (total immobility) "Abdominal" (abdominal vibrations) "Grooming" (passing a leg or a pedipalp between their chelicereae) "Leg-waving" (raising and lowering a leg of their first pair) "Standing" (standing on hind legs vertically to a wall) "Drumming" (drumming the ground with their first pair of legs)
Exploratory activity Distance covered (cm/min)	Time of mobility Number of arms visited

2.4. Locomotor and Exploratory Activity in a Novel Environment

Six experimental groups (*n* = 25 males and 25 females in each experimental group; Figure 1) were observed in two novel environments and each spider was only tested once: -NE group: female and male adults that had developed in their natural environment tested between 2 and 3 h after capture; -NE-15 group: female and male adults that had developed in their natural environment tested after 15 days in captivity; -L group: female and male adults that had developed in the laboratory in large terraria without any matter on the base; -LM group: female and male adults that had developed in the laboratory in large terraria with matter on the base; -S group: female and male adults that had developed in the laboratory into small terraria without any matter on the base; -SM group: female and male adults that had developed in the laboratory in small terraria with matter on the base. All subjects were observed at the same time of day (between 14.00 and 17.00 h), at 20 ± 1 °C with 40 ± 2% relative humidity in natural light (90 lx in each glass arena, measured by a Spengler Luxmeter before each behavioural test, Securimed, Cappelle-La-Grande, France).

Each spider was observed successively in an open-field glass arena (15 cm in diameter, 5 cm high covered with a glass plate) and in a radial-arm maze (Figure 2). The maze was an array of eight arms in opaque white PVC (L × l: 10 × 2.5 cm and 3 cm high) radiating from a central starting area (6 cm in diameter) and covered with a transparent glass plate (L × l: 22 × 24 cm). Two arenas and two mazes were used simultaneously during a test so we could pair subjects in relation to sex and experimental environment, combining all possibilities (L/LM, L/S, L/SM, L/NE, L/NE-15, LM/S, LM/SM, LM/NE, LM/NE-15, S/SM, S/NE, S/NE-15, SM/NE, SM//NE-15, NE/NE-15). This methodological precaution was taken to counteract effects of possible changes in temperature and humidity on spiders' behaviour. To eliminate biases due to the positions of the arenas, I alternated the experimental groups between sides and between trials. The arenas were washed with acetone-water (5%) then ethanol (70%) between trials to eliminate any possible intraspecific cues (odour and dragline cues left by the previous spider) that could influence a spider's activity. They were reused 1 h after evaporation of washing solvents.

A   B

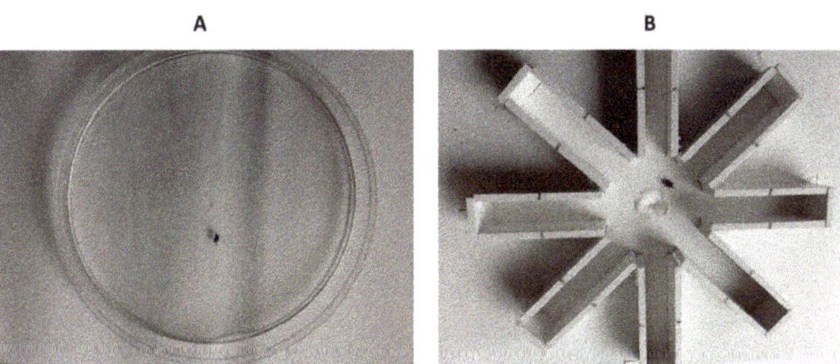

Figure 2. The open-field glass arena (**A**) and the eight radial-arm maze (**B**) used successively during behavioural tests for wild caught spiders immediately after capture (NE) or after 15 days in captivity (NE-15) and for laboratory reared spiders (L, LM, S and SM groups).

Spiders were observed individually in the arena and their behaviour and locomotor activities were recorded and analysed applying the method followed by Ruhland et al. [21]. The coordinates of spiders in the novel environment (arena and maze) were recorded every 12 frames using a Canon HD (HG20) camera and an automated video-based, digital-data collection system (Swiss-Track software 4.0.0 with the nearest neighbour tracking method).

After being weighed, a spider was placed in the centre of the arena under a bell (1.5 cm in diameters) 1 min prior to the test. The bell was removed, and the subject was allowed at least 2 min to acclimate to the arena prior to data collection. After the acclimation period, the spider's activity (locomotors activity and spontaneous activity without changing place; Table 1) was monitored for 10 min. The software program calculated a spider's locomotor activity directly (expressed in surface explored, cm/min). Other movements in the arena were recorded directly by the experimenter as indicated above (Table 1).

At the end of arena test, the spider was rapidly transferred into a vial and immediately placed in the central box of the radial arm maze under a bell (1.5 cm diameters) for evaluation of the exploratory activity. The bell was removed, and the spider was allowed at least 2 min to acclimate to the arena prior to data collection. After the acclimation period, the spider's exploratory activity in the maze was monitored during 10 min (Table 1). The software program calculated exploratory activity (time of mobility, number of arms visited) expressed in distance covered (cm/min) of each spider directly. When a distance covered was less than 3 mm/second it was not taken into account.

2.5. Statistical Analyses

Statistical analyses were performed using STATISTICA 6.0 for WINDOWS (Statsoft Inc., Tulsa, OK, USA). A generalised linear mixed test (GLMM, negative binomial distribution) was conducted with data of body mass, the surface explored and distance covered after transformation of data in log to normalize the data. Frequencies of spontaneous behaviours were divided by 100 and then converted to $\sqrt{}$arcsine proportions to normalize the data before a generalised linear mixed test was conducted. When differences among means were significant at the $p < 0.05$ levels, an a posteriori Wald test was used to establish inter-group comparisons. Means are given ±95% confidence intervals (CI).

3. Results

3.1. Body Mass

Body mass (Figure 3) differed significantly between spider groups and between sexes (environment effect: 4.095, df = 5, $p < 0.001$; sex effect: 20.886, df = 1, $p < 0.001$; environment*sex: 0.349, df = 5, $p = 0.083$). Body mass of NE females was significantly less than that of NE-15 females and laboratory reared females (Figure 3A). Body mass of LM females was significantly greater than that of all the other females. All males were significantly lighter than females and body mass of NE males was significantly less than that of NE-15 males and of males of all the laboratory-reared groups (Figure 3B).

3.2. Behaviour and Locomotor Activity of NE Spiders

Pardosa saltans moved about actively in their terrarium (Figure 4) and locomotor activity of the natural spider group varied significantly with time and between sexes (day effect: $\chi = 128.105$, df = 2, $p < 0.001$; sex effect: $\chi = 4.289$, df = 1, $p = 0.038$; day*sex: $\chi = 12.558$, df = 2, $p = 0.002$). Females and males after capture (0 day in captivity) mover significantly more than did the other. After 7 then 14 days in captivity, the locomotor activity of all the natural spiders decreased significantly and the locomotor activity levels of females and males became similar after 14 days in captivity.

The frequency of different behaviours (Figure 5) varied significantly between sexes and with time (sex effect: $\chi = 6.61$ to 7.47, df = 1, $p = 0.010$ to 0.006; day effect: $\chi = 8.82$ to 1057.42, df = 2, $p = 0.012$ to < 0.001; day*sex: $\chi = 17.97$ to 273.42, df = 2, $p < 0.001$). Males were significantly more active than females. Females and males significantly increased "grooming" and decreased "standing" after 14 days in captivity. As time since collection increased males became more inactive and increased "grooming" and "leg-waving" activities.

3.3. Behaviour in a Novel Environment (Open-Field Arena and Arm-Maze)

Levels of behavioural activities in a novel environment (open-field arena) varied significantly between spider groups (Figure 6) and between sexes (environment effect: $\chi = 46.21$ to 1924.87, df = 5, $p < 0.001$; sex effect: $\chi = 20.50$ to 162.82, df = 1, $p < 0.001$; environment*sex: $\chi = 14.25$ to 1075.06, df = 5, $p = 0.010$ to < 0.001). Only the natural spider group manifested "leg-waving" and only natural males manifested "drumming" activity. The natural females presented significantly more "standing" and less "grooming" than did the females of all the laboratory groups. The natural males manifested significantly less "grooming" and "inactivity", and more "standing" than the other males. Levels of spontaneous behavioural activity without changing position did not vary significantly between the laboratory female and male groups.

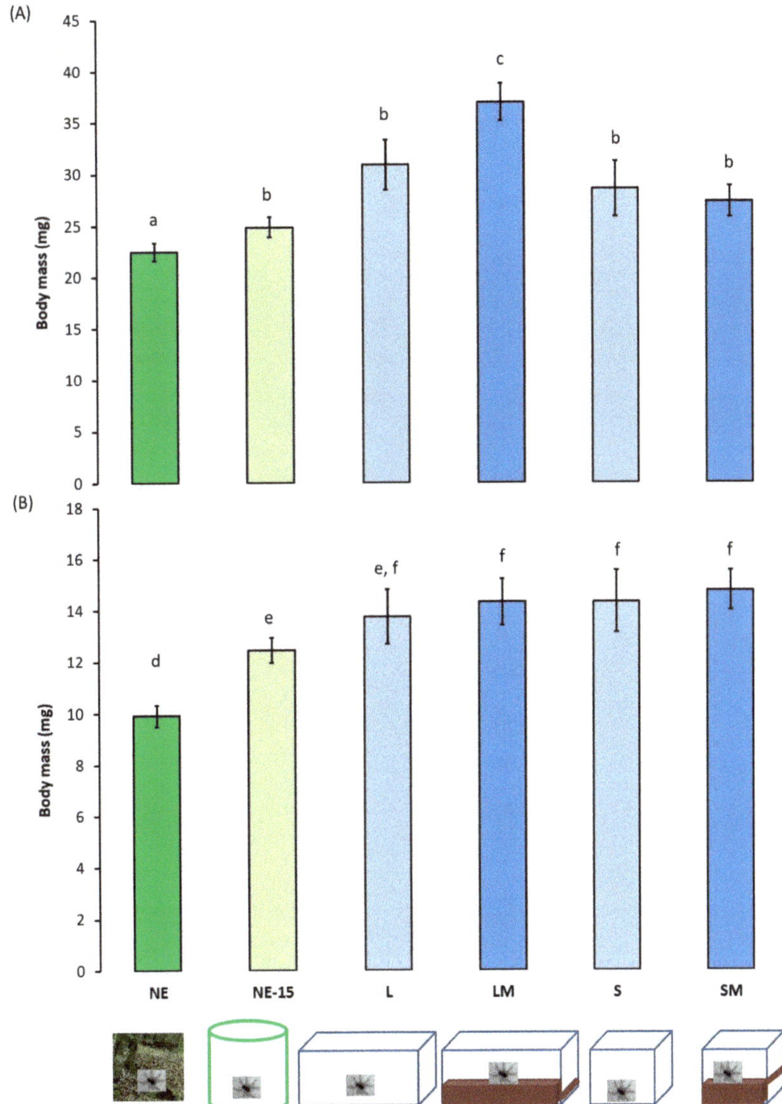

Figure 3. Body mass ±95% confidence intervals (in mg) of adult female (**A**) and male (**B**) *Pardosa saltans* in relation to the environment where they developed: in their natural environment and tested between 2 and 3 h after capture (NE); in their natural environment and tested after 15 days in captivity (NE-15); in the laboratory in large terraria without matter on the base (L); in the laboratory in large terraria with matter on the base (LM); in the laboratory in small terraria without matter on the base (S); in the laboratory in small terraria with matter on the base (SM). Data were compared using generalised linear mixed mod (GLMM, negative binomial distribution). Different letters indicate a significant difference at $p < 0.050$, post hoc Wald test.

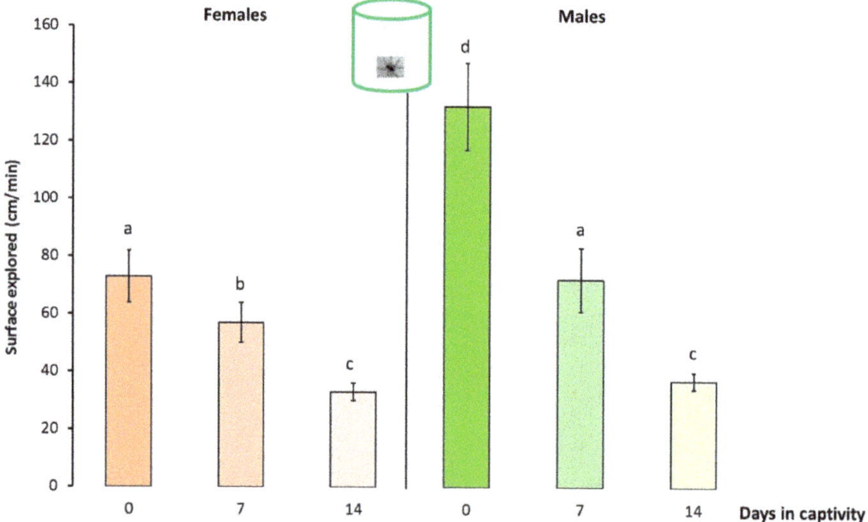

Figure 4. Locomotor activity ±95% confidence intervals (expressed in surface explored, cm/min) of adult female and male *Pardosa saltans* in terraria after capture in their natural environment (0 day in captivity), seven days after capture (7 days in captivity) and 14 days after capture (14 days in captivity). Data were compared using generalised linear mixed mod (GLMM, negative binomial distribution). Different letters indicate a significant difference ($p < 0.050$, post hoc Wald test).

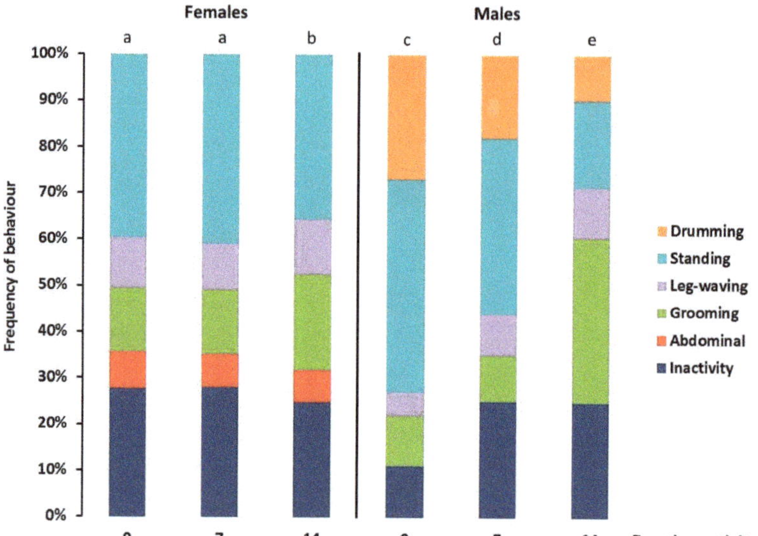

Figure 5. Spontaneous behavioural activities without changing position of adult female and male *Pardosa saltans* (in frequencies) in terraria after capture in their natural environment (0 day in captivity), seven days after capture (7 days in captivity) and 14 days after capture (14 days in captivity). Frequencies of behavioural events were compared using generalised linear mixed mod (GLMM, negative binomial distribution). Different letters indicate a significant difference ($p < 0.050$, post hoc Wald test).

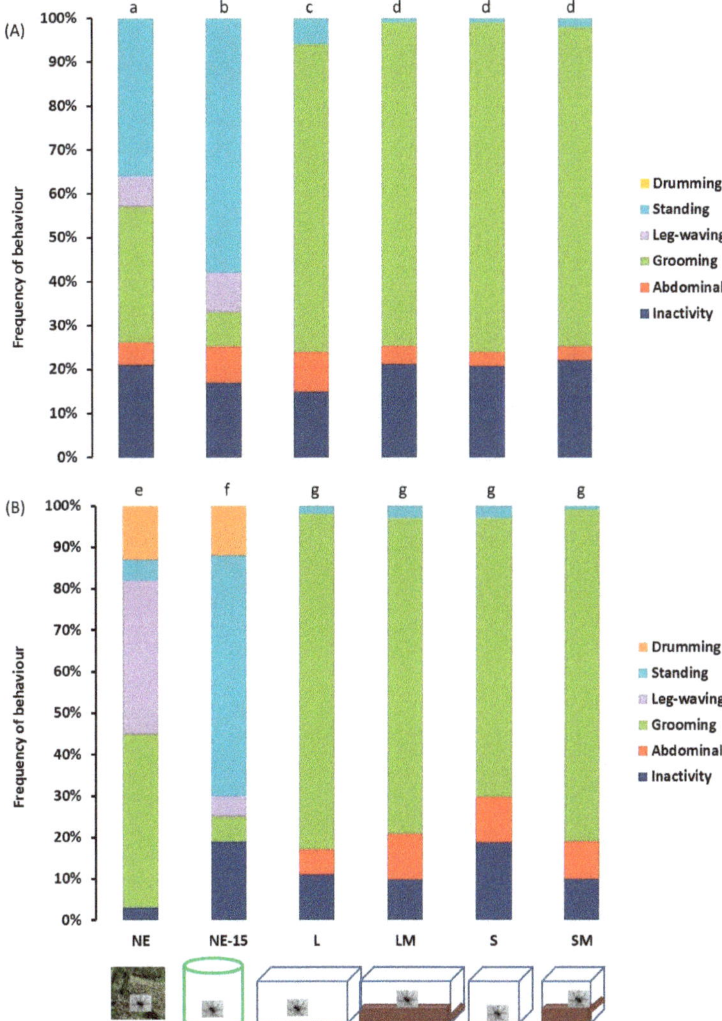

Figure 6. Spontaneous behavioural activities without changing position of adult female (**A**) and male (**B**) *Pardosa saltans* (in frequencies) in the open-field arena in relation to the environment where they developed: (NE) in their natural environment and tested 2–3 h after capture; (NE-15) in their natural environment and tested after 15 days in captivity; (L) in the laboratory in large terraria without matter on the base; (LM) in the laboratory in large terraria with matter on the base; (S) in the laboratory in small terraria without matter on the base; (SM) in the laboratory in small terraria with matter on the base. Frequencies of behaviours were compared using generalised linear mixed mod (GLMM, negative binomial distribution). Different letters indicate a significant difference ($p < 0.050$, post hoc Wald test).

3.4. Locomotor and Exploratory Activities in a Novel Environment

All spiders walked preferentially close to the periphery of the open-field arena and radial arm maze. Surfaces and distances explored during locomotor activity in the open-field arena and the radial arm maze (Figures 7 and 8) varied significantly between the experimental spiders groups (environment effect: $\chi = 13.473$, df = 5, $p < 0.001$; sex effect: $\chi = 2.046$, df = 1, $p = 0.155$; environment*sex: $\chi = 0.264$, df = 5, $p = 0.932$ for surfaces explored; environment effect: $\chi = 13.325$, df = 5, $p < 0.001$; sex effect: $\chi = 8.386$ df = 1, $p = 0.004$; environment*sex: $\chi = 3.399$, df = 5, $p = 0.006$ for distance covered). Thus, NE females and males moved about more: they explored larger surfaces and covered greater distances. After 15 days in captivity the surfaces explored by females and males did not differ significantly from the surfaces explored by LM and SM females. The surface explored and distances covered by the L and S spiders were significantly lower.

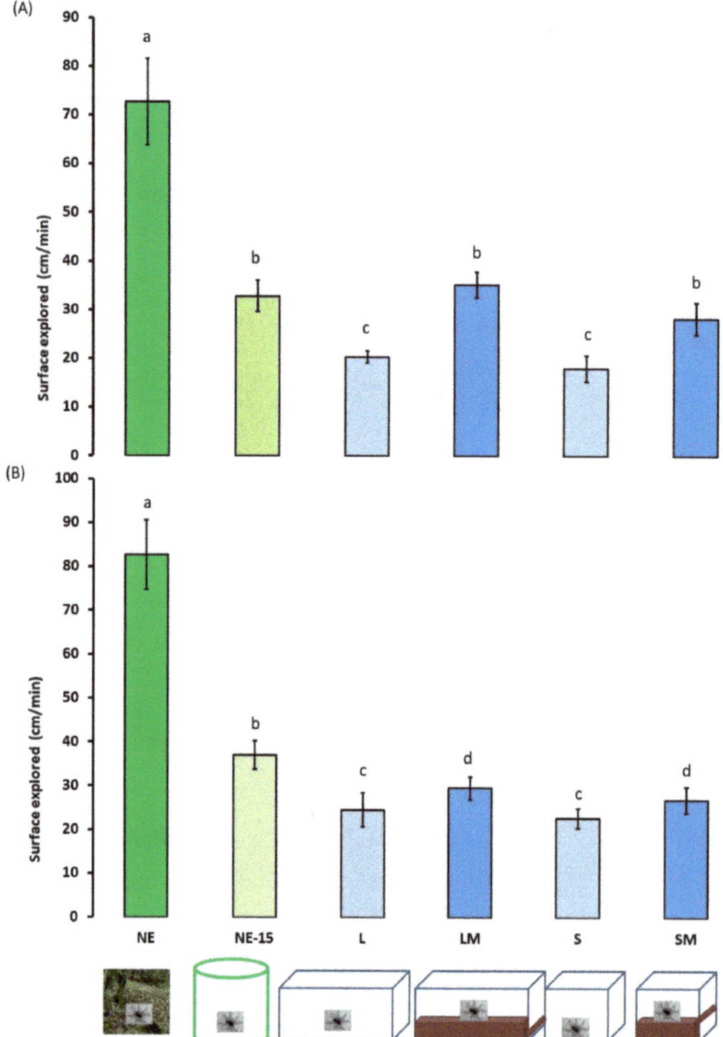

Figure 7. Locomotor activity $\pm 95\%$ confidence intervals (expressed in surfaces explored, cm/min) of adult female (**A**) and male (**B**) *Pardosa saltans* in a open-field arena in relation to the environment where they developed: (NE) in their natural environment and tested 2–3 h after capture; (NE-15) in their natural environment and tested after 15 days in captivity; (L) in the laboratory in large terraria without matter on the base; (LM) in the laboratory in large terraria with matter on the base; (S) in the laboratory in small terraria without matter on the base; (SM) in the laboratory in small terraria with matter on the base. Data were compared using generalised linear mixed mod (GLMM, negative binomial distribution). Different letters indicate a significant difference ($p < 0.050$, post hoc Wald test).

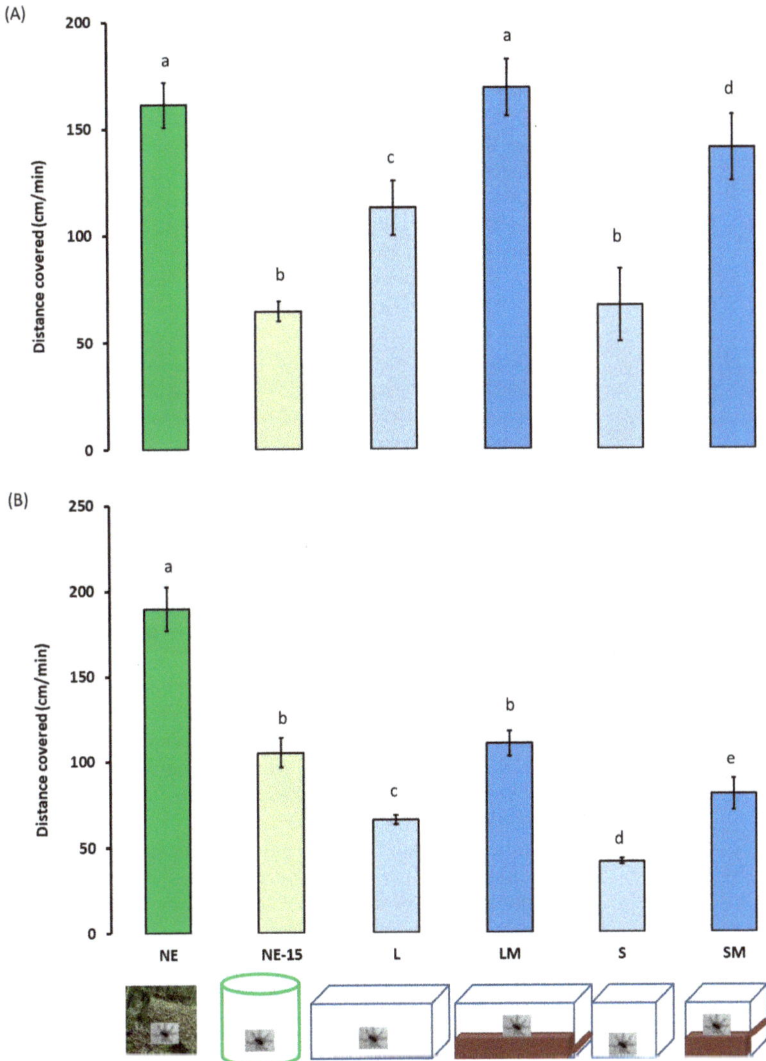

Figure 8. Exploratory activity ±95% confidence intervals (expressed in distances covered, cm/min) of adult female (**A**) and male (**B**) *Pardosa saltans* in a radial arm maze in relation to the environment where they developed: (NE) in their natural environment and tested 2–3 h after capture; (NE-15) in their natural environment and tested after 15 days in captivity; (L) in the laboratory in large terraria without matter on the base; (LM) in the laboratory in large terraria with matter on the base; (S) in the laboratory in small terraria without matter on the base; (SM) in the laboratory in small terraria with matter on the base. Data were compared using generalised linear mixed mod (GLMM, negative binomial distribution). Different letters indicate a significant difference ($p < 0.050$, post hoc Wald test).

4. Discussion

In this study, I used a combination of behavioural assays to examine the effects of rearing conditions on the locomotor, exploratory, and spontaneous behavioural activity of the wolf spiders, *P. saltans*. My objective was to answer mainly three questions: does captivity, after only 15 days, affect the locomotor and exploratory behaviour of adult spiders developed in their natural environment? Does captivity affect female and male adults differently? Should wandering spiders be raised and maintained in large or small terraria, with or without litter? My results show that activity levels varied based on sex and treatment.

When introduced into a novel environment immediately after capture, my results show that adult *P. saltans* spiders developed in the wild exhibit high levels of locomotor and exploratory activity. Males are twice as active as females as previously observed in other studies with other invertebrate species [8–11,22]. Thus, males explore larger surfaces but contrary to our predictions they do not cover significantly longer distance than the females. During phases of immobility, spiders exhibit six spontaneous behaviours: "inactivity", "abdominal vibration", "grooming", "leg-waving", "standing" and "drumming". "Grooming" of the legs and pedipalps are the most frequent activities observed. These activities can be explained by the fact that adult spiders possess many chemoreceptors (receptors for chemical signals) and mechanoreceptors (receptors for vibratory signals) on the tarsi of their legs and their pedipalps [23,24]. These sensory receptors are highly stressed when travelling in heterogeneous environments and individuals must regularly remove, by "grooming", any artefacts that could impair the perception of communication signals. "Grooming" activities increase the reception of communication signals when an individual is foraging or sexually active, especially in a novel environment. In addition, males mainly manifest "leg-waving" and "drumming" activities, characteristically emitted by adult male spiders during sexual activities when searching for a sexual partner in their natural environment [8,16]. Females placed in a novel environment mainly perform "standing" against the walls of the experimental enclosures.

Captivity results in rapid behavioural acclimatization of spiders that had developed in their natural environment. Thus after 7 days of captivity, the locomotor and exploratory activities of both female and males spiders decreased significantly. After 15 days in captivity, these spiders moved significantly less in their terraria and the levels of locomotor activity of females and males during behavioural assays were comparable to those of individuals reared in the laboratory in small terraria without substrate. These results corroborate observations of a salticid spider *Phidippus autax* maintained under artificial conditions for 4 months [15]. This decrease was not correlated with a change in ambient temperature and/or humidity at the time of testing, unlike for other animals [25–27]. This decrease was also not correlated with a lack of energy as these female and male spiders had put on weight. Consequently, the decrease in activity shown by spiders from the natural environment is clearly linked to an effect of captivity.

Captivity also results in a change in the spontaneous behavioural activities exhibited by the spiders developed in their natural environment as observed for other animal species [2,19]. It is the males who are the most affected by captivity. Thus after only 7 days of captivity, the locomotor activity of males is comparable to that of females. This gradual decline is accompanied by a decrease of "standing" and "drumming" activities and, conversely, by an increase in "grooming" and "inactivity". This change in behavioural activities of females is less visible, although they, like males, show a significant decrease in their locomotor activity after 14 days in captivity. Thus, a change of environment in adulthood leads *P. saltans* to gradually, significantly but quickly modify its locomotor and exploratory activities. This effect is perhaps linked to the sudden modification of their environment, which induces greater deprivation for individuals accustomed to moving freely over large surfaces and in a complex physical environment. Thus, my study shows that being kept in a terrarium induces a progressive and negative modification of the behavioural activities of *P. saltans* spiders, as observed for three species of tarantulas [28]. In their natural environment, *P. saltans* moves in different environments such as undergrowth and meadows. These spiders are therefore able to adapt quickly to a novel environment, which explains the rapid change in their behaviour the days following their collection.

With regard to effects of breeding on spiders, my results show that the size of the terraria did not appear to affect the spontaneous behaviour or the locomotor activity of adult spiders: female and male spiders reared in either small or large terraria exhibit the same spontaneous behaviours and the same levels of surface explored contrary to observations by other researchers on Salticid spiders [15,18,19]. However, the size of the terraria affects their exploratory activity: spiders, and more particularly males, reared in small terraria cover shorter distances when exploring a novel environment. These differences in activity are once again not linked to abiotic factors (temperature, humidity) or to the weight of the spiders. Indeed, the spiders bred in the laboratory and according to sex have comparable masses.

Conversely, the presence of litter in terraria promotes the levels of *P. saltans* spiders' locomotor and exploration activities as for *Marpissa muscosa* spiders [6]. Thus, females and males reared in terraria with litter are more active, cover greater areas and distances than spiders reared in terraria without litter. These results show that the presence of substrate stimulates favourably the locomotor and exploratory activities of wandering spiders during their development as for vertebrates [29,30]. However, spiders that have developed in their natural environment are generally more active, move and explore more than those that have been reared in the laboratory even in the presence of litter. In addition, males reared in the laboratory, even in the presence of litter in their terrarium, show low levels of locomotor and exploratory activities comparable to those of females. My maintenance conditions in an artificial environment are too restrictive for the simple fact of adding a substrate or using a larger terrarium to be sufficient to compensate deprivations related to these breeding methods, in particular for males.

Furthermore, the spontaneous activities exhibited by the spiders reared in the laboratory during the behavioural tests show homogeneity between the different experimental groups. Neither the size nor the presence of litter influences the behavioural activity of the spiders that had developed in captivity for seven months. These female and male spiders exhibit "grooming" and "abdominal" vibrations when exploring a novel environment. The spiders reared in the laboratory show no "drumming" or "leg-waving" activities. These activities are used by adult spiders during the search for a sexual partner. These results suggest that these behavioural activities may emerge only as a consequence of being reared in a social environment, as observed for *Brachypelma smithi* tarantulas [19]. In addition, in their natural environment, spiders regularly encounter conspecifics. These encounters elicit the manifestation of behavioural characteristics of both their species and sex such as "leg-waving" and "drumming". The spiders in the laboratory group had been reared in social isolation during the seven months their development lasts, that is, they had never been in contact with any conspecifics. This would explain why female and male spiders reared in the laboratory are less active and do not "leg-waving" or "drumming". The absence of these behaviours indicates that these activities are linked to social learning that takes place during development, and that social isolation induces a reduction in behavioural complexity as observed for vertebrates [2,31–34]. Additionally, this socially deprived environment probably affects learning abilities and the central nervous system of *P. saltans* as observed for the jumping spider, *Marpissa muscosa* [35,36].

On the other hand, it is interesting to note that the body mass of *Pardosa saltans* differs according to sex and environment. Males weighed less than females, confirming previous studies of these spiders [37], and the spiders collected in the wild were lighter than those that have been raised for seven months in the laboratory in a controlled environment. After 15 days in captivity, rearing conditions affected positively their weight gain. Furthermore, males reared in the laboratory weighed less than laboratory reared females. However, under laboratory conditions, the males were fed ad libitum and maintained under relatively stable abiotic conditions during development (e.g., no rain, no frost). Under these conditions, one might think that these males would probably spend less energy to fight heat loss and to explore their environment in search of food compared to other males subjected to environmental constraints, and therefore they would be able to store more energetic stock in their organism. My results showed that this is not the case, so I hypothesised that inter-individual mass variations are influenced more by genetic factors than by biotic or abiotic environmental factors. The fact that my results concerning females differ from those for males suggests that regulation of weight gain differs in relation to sex. Indeed, the body weights of females varied according to their living environment. Females from the natural environment weighed less than females reared in the laboratory in a controlled environment and fed ad libitum. In addition, females reared in large terraria weighed more than females living in small terraria. *Pardosa saltans* females therefore exhibit morphological plasticity allowing individuals to adapt to the spatial characteristics of their captive environment when their diet is optimal. This result can be linked to observations of other animal species kept in controlled environments (e.g., fish in aquariums: [38]; lobsters: [39]). An abiotic factor such as the volume of the rearing environment and the presence of a litter therefore plays a role in the development of the body of female spiders.

5. Conclusions

My results show that, as for other animal species, abiotic and biotic living conditions influence the locomotor and exploratory behaviours activities of the wandering spider *P. saltans*. The rearing conditions I used influenced positively their foraging behaviour (weight gain) but may have influenced profoundly and negatively the behaviour of adults. My study also highlights the fact that their environment impacts males and females differently. Thus, the modifications of female spiders' behaviour can be attenuated by physical enrichment of their environment. In contrast, males were more impacted by my laboratory conditions, and simply adding substrate was not sufficient to reduce this impact. These negative changes of *P. saltans'* behaviour occurred rapidly (in less than 15 days). These results underscore the importance of taking into account the length of time individuals have been kept in the laboratory before drawing conclusions behavioural studies concerning animal well-being.

Funding: This research received no external funding.

Institutional Review Board Statement: Not applicable.

Data Availability Statement: The data presented in this study are available on request from the corresponding author.

Acknowledgments: I would like to thank Maxime Hervé for blind treatment analyses of the behavioural experiments; Ann Cloarec for reading the manuscript.

Conflicts of Interest: The author declares no conflict of interest.

References

1. Baumans, V. Environmental enrichment for laboratory rodents and rabbits: Requirements of rodents, rabbits, and research. *ILAR J.* **2005**, *46*, 162–170. [CrossRef]
2. Morgan, K.N.; Tromborg, C.T. Sources of stress in captivity. *Appl. Anim. Behav. Sci.* **2006**, *102*, 262–302. [CrossRef]
3. Clubb, R.; Mason, G. Captivity effects on wide-ranging carnivores. *Nature* **2003**, *425*, 473–474. [CrossRef] [PubMed]
4. Clubb, R.; Mason, G. Natural behavioural biology as a risk factor in carnivore welfare: How analysing species differences could help zoos improve enclosures. *Appl. Anim. Behav. Sci.* **2007**, *102*, 303–328. [CrossRef]
5. Block, W.; Baust, J.G.; Franks, F.; Johnston, I.A.; Bale, J. Cold tolerance of insects and other arthropods and discussion. *Phil. Trans. R. Soc. Lond. B* **1990**, *326*, 613–633.
6. Liedtke, J.; Redekop, D.; Schneider, J.M.; Schuett, W. Early environmental conditions shape personality types in a jumping spider. *Front. Ecol. Evol.* **2015**, *3*, 134. [CrossRef]
7. Punzo, F.; Alvarez, J. Effects of early contact with maternal parent on locomotor activity and exploratory behavior in spiderlings of *Hogna carolinensis* (Araneae: Lycosidae). *J. Insect Behav.* **2002**, *15*, 455–465. [CrossRef]
8. Persons, M.H. Hunger effects on foraging responses to perceptual cues in immature and adult wolf spiders (Lycosidae). *Anim. Behav.* **1999**, *57*, 81–88. [CrossRef]
9. Gerlach, A.; Voigtländer, K.; Heidger, C.M. Influences of the behaviour of epigenic arthropods (Diplopoda, Chilopoda, Carabidae) on the efficiency of pitfall trapping. *Soil Org.* **2009**, *81*, 773–790.
10. Jensen, C.S.; Garsdal, L.; Baatrup, E. Acetylcholinesterase inhibition and altered locomotor behavior in the carabid beetle *Pterostichus cupreus*. A linkage between biomarkers at two levels of biological complexity. *Environ. Toxicol. Chem.* **1997**, *16*, 1727–1732. [CrossRef]

11. Bayley, M. Prolonged effects of the insecticide dimethoate on locomotor behaviour in the woodlouse, *Porcellio scaber* Latr. (Isopoda). *Ecotoxicology* **1995**, *4*, 79–90. [CrossRef] [PubMed]
12. Babu, K.S. Post embryonic development of the central nervous system of the spider *Argiope aurantia* (Lucas). *J. Morphol.* **1975**, *146*, 325–342. [CrossRef]
13. Baatrup, E.; Bayley, M. Quantitative analysis of spider locomotion employing computer-automated video-tracking. *Physiol. Behav.* **1993**, *54*, 83–90. [CrossRef]
14. Carere, C.; Wood, J.B.; Mather, J. Species differences in captivity: Where are the invertebrates? *Trends Ecol. Evol.* **2011**, *26*, 211. [CrossRef] [PubMed]
15. Carducci, J.P.; Jakob, E.M. Rearing environment affects behaviour of jumping spiders. *Anim. Behav.* **2000**, *59*, 39–46. [CrossRef]
16. Gibson, J.S.; Uetz, G.W. Effect of rearing environment and food availability on seismic signalling in male wolf spiders (Araneae: Lycosidae). *Anim. Behav.* **2012**, *84*, 85–92. [CrossRef]
17. Tarsitano, M.S.; Jackson, R.R. Jumping spiders make predatory detours requiring movement away from prey. *Behaviour* **1994**, *131*, 65–73. [CrossRef]
18. Folz, H. Effects of Complexity of Rearing Environment on Activity Level and Foraging Behavior of *Hogna helluo* (Araneae, Lycosidae). Bachelor's Thesis, Miami University, Oxford, OH, USA, 2005.
19. Bengston, S.E.; Pruitt, J.N.; Riechert, S.E. Differences in environmental enrichment generate contrasting behavioural syndromes in a basal spider lineage. *Anim. Behav.* **2014**, *93*, 105–110. [CrossRef]
20. Tarsitano, M. Route selection by a jumping spider (*Portia labiata*) during the locomotory phase of a detour. *Anim. Behav.* **2006**, *72*, 1437–1442. [CrossRef]
21. Ruhland, F.; Chiara, V.; Trabalon, M. Age and egg-sac loss determine maternal behaviour and locomotor activity of wolf spiders (Araneae, Lycosidae). *Behav. Process.* **2016**, *132*, 57–65. [CrossRef]
22. Töpfer-Hofmann, G.; Cordes, D.; von Helversen, O. Cryptic species and behavioural isolation in the *Pardosa lugubris* group (Araneae, Lycosidae), with description of two new species. *Bull. Br. Arachnol. Soc.* **2000**, *11*, 257–274.
23. Gallo, N.D.; Jeffery, W.R. Evolution of space dependent growth in the teleost *Astyanax mexicanus*. *PLoS ONE* **2012**, *7*, e41443. [CrossRef] [PubMed]
24. Aiken, D.E.; Waddy, S. Space, density and growth of the lobster (Homarus americanus). In Proceedings of the Annual Meeting—World Mariculture Society; Wiley-Blackwell: Oxford, UK, 1978; pp. 459–467.
25. Schmitt, A.; Schuster, M.; Barth, F.G. Daily locomotor activity patterns in three species of *Cupiennius* (Araneae, Ctenidae): The males are the wandering spiders. *J. Arachnol.* **1990**, *18*, 249–255.
26. Barth, F.G. Spider senses—Technical perfection and biology. *Zoology* **2002**, *105*, 271–285. [CrossRef] [PubMed]
27. Trabalon, M. Chemical communication and contact cuticular compounds in spiders. In *Spider Ecophysiology*; Nentwig, W., Ed.; Spinger: Berlin/Heidelberg, Germany, 2013; pp. 125–140.
28. Bennie, M.; Loaring, C.; Trim, S. Laboratory husbandry of arboreal tarantulas (Theraphosidae) and evaluation of environmental enrichment. *Anim. Technol.* **2011**, *10*, 163–169.
29. Genaro, G.; Schmidek, W.; Franci, C.R. Social condition affects hormone secretion and exploratory behavior in rats. *Braz. J. Med. Biol. Res.* **2004**, *37*, 833–840. [CrossRef]
30. Herskin, M.S.; Jensen, K.H. Effects of different degrees of social isolation on the behaviour of weaned piglets kept for experimental purposes. *Anim. Welf.* **2000**, *9*, 237–249.
31. Patterson-Kane, E.G.; Hunt, M.; Harper, D. Rats demand social contact. *Anim. Welf.* **2002**, *11*, 327–332.
32. Weiss, I.C.; Pryce, C.R.; Jongen-Relo, A.L.; Nanz-Bahr, N.I.; Feldon, J. Effect of social isolation on stress-related behavioural and neuroendocrine state in the rat. *Behav. Brain Res.* **2004**, *152*, 279–295. [CrossRef]
33. Liedtk, J.; Schneider, J.M. Social makes smart: Rearing conditions affect learning and social behaviour in jumping spiders. *Anim. Cogn.* **2017**, *20*, 1093–1106. [CrossRef]
34. Steinhoff, P.O.M.; Liedtke, J.; Sombke, A.; Schneider, J.M.; Uhl, G. Early environmental conditions affect the volume of higher-order brain centers in a jumping spider. *J. Zool.* **2018**, *304*, 182–192. [CrossRef]
35. Anderson, J.F. Metabolic rates of spiders. *Comp. Biochem. Physiol.* **1970**, *33*, 51–72. [CrossRef]
36. Rott, A.S.; Ponsonby, D.J. The effects of temperature, relative humidity and hostplant on the behaviour of *Stethorus punctillum* as apredator of the two-spotted spider mite, *Tetranychus urticae*. *BioControl* **2000**, *45*, 155–164. [CrossRef]
37. Danks, H.V. The elements of seasonal adaptations in insects. *Can. Entomol.* **2007**, *139*, 1–44. [CrossRef]
38. Hansen, L.T.; Berthelsen, H. The effect of environmental enrichment on the behaviour of caged rabbits (*Oryctolagus cuniculus*). *Appl. Anim. Behav. Sci.* **2000**, *68*, 163–178. [CrossRef]
39. Patterson-Kane, E.G. Shelter enrichment for rats. *J. Am. Assoc. Lab. Anim. Sci.* **2003**, *42*, 46–48.

Article

Spider Assemblages of Tree Trunks and Tree Branches in Three Developmental Phases of Primeval Oak–Lime–Hornbeam Forest in the Białowieża National Park

Marzena Stańska [1] and Tomasz Stański [2],*

[1] Institute of Biological Sciences, Faculty of Sciences, Siedlce University of Natural Sciences and Humanities, 08-110 Siedlce, Poland
[2] Faculty of Sciences, Siedlce University of Natural Sciences and Humanities, 08-110 Siedlce, Poland
* Correspondence: tomasz.stanski@uph.edu.pl

Simple Summary: At present, the only place in Europe where the full development cycle of forests takes place on a large scale is the Białowieża Forest, because in most other forests dead or dying trees are eliminated, so the terminal (decay) phase does not occur there. Studies of animal assemblages inhabiting different forest phases are scarce as well as studies of spiders inhabiting tree trunks and branches. In this study, we compare spider assemblages inhabiting the tree trunks and branches in the optimal, terminal and regeneration phases of a primeval oak–lime–hornbeam stand in terms of their abundance, species diversity and species richness. We did not find differences in the total spider species richness between the analysed phases. However, we found that species diversity of both foliage-dwelling and trunk-dwelling spider assemblages was higher in the terminal phase compared to the other phases, which may indicate that this phase offers the most diverse niches for spiders as a result of the significant disturbance in the forest stand structure. Our research contributes to the understanding of the functioning of natural ecosystems, which can be useful for responsible forest management.

Abstract: The study was conducted in the Białowieża Forest, which is the only place in Europe where the full development cycle of forests takes place on a large scale. The objective of this study was to compare spider assemblages inhabiting tree trunks and tree branches in the optimal, terminal and regeneration phases of a primeval oak–lime–hornbeam stand, in terms of their abundance, species diversity and species richness. Spiders of tree branches were sampled using a sweep net into which branches were shaken, while spiders inhabiting tree trunks were collected using traps made of corrugated cardboard placed around the trunks. The three analysed phases did not differ in terms of total species richness. We found that the species diversity of both foliage-dwelling and trunk-dwelling spider assemblages was higher in the terminal phase compared to other phases, which may indicate that the former phase offered the most diverse niches for spiders as a result of the significant disturbance in the stand structure. In addition, we found fewer spider individuals and species in individual samples collected on tree branches from a plot in the regeneration phase compared to the other phases, which may be a consequence of the structure of the stand in this phase (low canopy cover, lush herbaceous vegetation).

Keywords: arboreal spiders; the Białowieża Forest; primeval forest

Citation: Stańska, M.; Stański, T. Spider Assemblages of Tree Trunks and Tree Branches in Three Developmental Phases of Primeval Oak–Lime–Hornbeam Forest in the Białowieża National Park. *Insects* **2022**, *13*, 1115. https://doi.org/10.3390/insects13121115

Academic Editors: Thomas Hesselberg and Dumas Gálvez

Received: 31 October 2022
Accepted: 1 December 2022
Published: 3 December 2022

Publisher's Note: MDPI stays neutral with regard to jurisdictional claims in published maps and institutional affiliations.

1. Introduction

Trees, because of their large size and complex structure, provide many unique and important microhabitats (e.g., trunks, foliage, branches, cavities) for many groups of invertebrates, including spiders [1–4]. Despite this fact, the spider fauna of trees is a rare subject of research. Blick [5] estimated the knowledge of spiders inhabiting tree trunks in forests of Central Europe at 5% compared to that of spiders inhabiting the ground. There are also few studies on spiders inhabiting tree branches [6]. Furthermore, in many of these studies, material was collected from different parts of trees or their strata and analysed together because of the use of nonselective methods such as insecticide fogging [7,8]. This may lead to incorrect conclusions, as individual microhabitats on trees vary greatly in structure and microclimatic conditions, and thus the spider assemblages inhabiting them are likely to be different. In contrast to such studies, here we separately analysed two microhabitats on trees, tree trunks and tree branches, in relation to the forest stand development phase.

The present study was conducted in the Białowieża National Park, where valuable natural European lowland forests are preserved. These forests are characterised as a multispecies community of trees, with a multi-layered and unevenly aged stand structure, considerable tree heights and a large amount of dead wood [9,10]. Unlike most forests in Europe, a complete cycle of forest stand development takes place here [11,12]. Several developmental phases can be distinguished in that cycle; however, their number is a matter of dispute. For example, Miścicki [13] defined eight phases (initial, juvenile, even-aged pole, premature, optimal, terminal, decay, regeneration), whereas Bobiec et al. [11] distinguished six phases (regeneration, young, pole, late pole, optimal and terminal). In our study, we included three of these phases, optimal, terminal/decay and regeneration, which are relatively easy to distinguish because of the significant differences in their stand structure. It is worth emphasising, however, that the decay (terminal) phase does not occur in most European forests as a result of logging and the elimination of dying trees.

Changes in invertebrate assemblages during the development cycle of temperate forests have rarely been studied, and when they have, the studies involved monocultures or forest plantations [14–16]. In the Białowieża Forest, such studies were conducted by Trojan et al. [17] in pine stands and included 27 taxa of animals (including spiders). Moreover, Stańska and Stański [18] studied plant-dwelling spider assemblages in different developmental phases of a primeval oak–lime–hornbeam stand. This study contained only spiders inhabiting herbaceous vegetation, which is a completely different habitat than trees. To our knowledge, there are no other studies discussing this problem in primeval forests.

Spiders are an excellent model group with which to study the effects of changes in the structure of a forest on the animal assemblages that inhabit it. Their abundance, species richness and diversity are affected by such factors as tree species diversity, the type of forest, its structure and canopy openness [19–24].

The objectives of this study were: (1) to determine the species composition of spider assemblages on tree branches and tree trunks in optimal, terminal and regeneration phases of primeval oak–lime–hornbeam forest; (2) to compare spider assemblages between these phases of stand development in terms of spider abundance (adults and juveniles separately), species richness and species diversity; and (3) to assess how the number of individuals and the number of species have changed over time (particular sampling months).

Many studies have shown that structurally diverse habitats support high species diversity, species richness and an abundance of spiders because they provide a large number of niches and diverse microhabitats [25–28]. Therefore, we hypothesised that spider species richness, diversity and abundance would be the highest on a plot with the terminal phase where, on the one hand, significant habitat disturbance has occurred (broken branches, emerging canopy gaps) and, on the other hand, mature, standing trees are still present.

2. Materials and Methods

2.1. Study Site

The Białowieża Forest, located on the Polish–Belarusian border, is a remnant of forests that covered much of temperate Europe centuries ago. Most of the area in the Polish part is under forest management, but the most valuable forest stands are protected as the Białowieża National Park (hereafter BNP). Human activity here is limited to scientific research and guided tourist walks. Forest stands in the BNP may be considered primeval forests, as evidenced by their multi-layered and uneven-aged structure, multispecies tree community, significant tree heights and a large amount of dead wood [9,10]. In addition, forest stands in the BNP have a heterogeneous structure, which is manifested in the fact that different developmental stages or forest types occupy small areas next to each other [11].

Our study was conducted in an oak–lime–hornbeam stand, which is the most common forest type in the BNP. In each of the three developmental phases of the forest, optimal, terminal and regeneration, one study plot (20 × 40 m rectangle) was selected. Trees growing on the optimal phase plot (52°43′50″ N; 23° 51′40″ E) were characterised by good vitality and a large diameter at breast height, and their crowns formed a dense canopy (above 90% cover). The most common tree species in this developmental phase were European hornbeam *Carpinus betulus*, pedunculate oak *Quercus robur*, Norway spruce *Picea abies*, small-leaved lime *Tilia cordata* and Norway maple *Acer platanoides*. Trees on the plot (52°43′30″ N; 23°51′50″ E) with the forest stand in the terminal phase of development had a large diameter at breast height, but were usually in poor condition, as indicated by the presence of numerous dead branches and large fragments of decayed wood. Gaps in the canopy of the forest stand resulted from many large branches breaking off from the trunks (canopy cover of about 80%). The dominant tree species on this plot were European hornbeam, pedunculate oak and Norway spruce. The forest stand in the regeneration phase (52°43′10″ N; 23°51′00″ E) was characterised by the presence of patches without trees or with single trees as a result of strong winds that had felled most of the old trees 20 years before our research. Therefore, the canopy cover was very thin (about 20%), and lying deadwood was very abundant. In addition, there were a large number of young trees. The forest stand on this plot consisted mainly of European hornbeam, small-leaved lime, Norway spruce and pedunculate oak.

2.2. Data Collection

Spiders were collected from tree branches from April to November 2000. A total of ten samples were collected from each study plot: one sample in April, two samples in May, two samples in June, two samples in July, two samples in October and one sample in November. Spiders were collected from the branches of different trees, each time being selected randomly. The spiders belonged to different species, but the European hornbeam was sampled most frequently because this species had the easiest access to these branches (they were at the right height). On each sampling date, material was collected on each plot from ten branches of a similar size (1 × 0.5 m), located at a height of 1–2 m. The sampled branches were placed in the sweep net and then shaken vigorously, after which they were carefully inspected to collect spiders that had not fallen into the net. Because of the low abundance of spiders, the material from ten branches collected from each plot on each sampling date was combined into one sample.

Spiders on tree trunks were collected from June 1998 to October 2000 every month except November, December, January and February. Spiders were sampled using traps made of corrugated cardboard (25 cm wide), which were placed around trunks with their corrugated surface facing inwards. On each plot, five traps were placed on live trees (the same procedure was used throughout the study period) of a similar diameter (two on hornbeam, two on lime, one on spruce) at a height of 1.5 m above the ground. During sampling, the traps were removed from the trunks and the spiders sitting on them were collected. In addition, spiders that remained on the bark at a trap site were also collected. The material from five traps collected from each plot on each sampling date was pooled as one sample due to the low abundance of spiders. In total, the material was collected 18 times in the optimal and regeneration phase and 19 times in the terminal phase (during one control in the optimal phase and one control in the regeneration phase, some destroyed traps were found; thus, two samples were excluded from the analysis).

The collected spiders were preserved in 75% alcohol and then identified in the laboratory to the species level or, if this was not possible, as in the case of many juvenile specimens, to the higher taxon. The material was deposited at the Institute of Biological Sciences, Siedlce University of Natural Sciences and Humanities, Poland.

2.3. Statistical Analysis

To estimate sampling sufficiency on the study plots, richness estimators (Chao1, Chao2, Jackknife1, Jackknife2 and Michaelis–Menten) were calculated using 100 randomisations in EstimateS software version 9.1.0 [29]. To check whether the plots in different developmental phases differed in terms of species richness (i.e., the number of species recorded throughout the study period), rarefaction curves were calculated for the observed species richness with 95% confidence limits, based on the bootstrap method with 100 replications [30]. The species richness computed for each phase was considered significantly different when the confidence limits did not overlap [31,32].

The formula for the Shannon index (H′) was used to calculate the species diversity:

$$H' = - \Sigma \, pi \, \ln \, (pi)$$

where pi is the proportion of individuals of species i [33].

The Hutcheson test was used to compare Shannon diversity indices calculated for plots in different developmental phases using formulas prepared in Excel [34].

Generalised linear models (GLMs) were used to assess the association of the number of collected spider individuals and spider species with the developmental phase of the forest stand and the sampling period. In the models where the response variable was the number of collected spider species and the number of adult individuals, Gaussian error distribution and the identity link function were used. In the model where the response variable was the number of collected juvenile spider individuals, the Gaussian error distribution and the log-link function were used. The "developmental phase" and "sampling month" were treated as fixed categorical explanatory variables. If a given variable showed a significant effect in a model, paired contrasts were calculated to find significant differences between its levels. These calculations were performed in SPSS 21.0 for Windows.

3. Results

3.1. Spiders of Tree Branches

A total of 725 spider individuals from eight families were collected on tree branches during the study period (320 individuals in the optimal phase, 236 individuals in the terminal phase and 169 individuals in the regeneration phase). Juvenile spiders dominated in the collected material in each developmental phase (655 individuals in total, ca. 90%). A total of 591 individuals were identified to the species level: 266 from the optimal phase plot, 188 from the terminal phase plot and 137 from the regeneration phase plot. A total 24 species were identified (17 in the optimal phase, 16 in the terminal phase and 11 in the regeneration phase), of which 8 were common to all plots. A total of 13 species were represented by only 1 individual captured on a given plot (Table 1). However, the calculated estimators indicated much higher species richness, especially for the optimal phase plot and the terminal phase plot (Table 2).

Trematocephalus cristatus was the most abundant species both in the optimal phase (where it accounted for 41.7% of the individuals identified to the species level) and the regeneration phase (37.2%), while *Neriene peltata* was most abundant in the terminal phase (25.5%), although the proportion of the former species was only slightly lower in this case (Table 1). The analysis of rarefaction curves revealed that the three studied developmental phases did not differ from each other in terms of the total species richness found on tree branches (Figure 1).

Table 1. Spiders collected on tree branches in three developmental phases of oak–lime–hornbeam stands in the Białowieża National Park (spider families, genus and species in alphabetical order). The percentages presented in parentheses, next to the number of individuals, show the proportion of each species. All individuals identified to the species level were included in the percentage composition, but values are only shown for species that reached at least 5%. Abbreviations: Ad./Juv.— number of adult/juvenile spider individuals, un.—individuals identified only to the family level. Roman letters indicate the months in which a given species was recorded.

Family/Genus/Species	Optimal Phase		Terminal Phase		Regeneration Phase	
	Ad./Juv.	Months	Ad./Juv.	Months	Ad./Juv.	Months
Anyphaenidae						
Anyphaena accentuata	(12%) -/32	IV–VII, X, XI	(20%) 1/37	IV–VII, X, XI	(36%) 1/49	IV–VII, X, XI
Araneidae						
Araneus diadematus			-/1	VII		
Araniella sp.	-/1				-/5	
Cyclosa conica	4/5	IV, V, VII, X	1/7	IV, V, X, XI	2/-	IV
Clubionidae						
Clubiona sp.	-/2				-/2	
Linyphidae						
Diplocephalus picinus			1/-	VI		
Entelecara acuminata	1/-	VII	1/-	VII		
Helophora insignis	-/5	VII	-/2	VII		
Linyphia triangularis	1/-	VII				
Linyphiidae un.	-/6		-/12		-/3	
Neriene clathrata	-/1	VII				
Neriene emphana	(6%) 2/15	IV, VI, VII	2/7	IV–VII, XI	-/1	V
Neriene montana	-/1	X	-/4	VII, X	1/-	V
Neriene peltata	(6%) 9/7	IV, V, XI	(26%) 6/42	IV–VI, X, XI	(8%) -/11	X
Neriene sp.	-/20		-/3		-/1	
Pityohyphantes phrygianus			-/1	X		
Porrhomma pygmaeum	5/-	IV			3/-	V
Tapinocyba insecta	1/-	IV				
Tapinocyba pallens			1/-	IV		
Trematocephalus cristatus	(42%) -/111	V, X, XI	(22%) -/42	X, XI	(37%) 1/50	IV, V, X, XI
Philodromidae						
Philodromus dispar	1/-	V				
Philodromus sp.	-/7		-/6		-/1	
Tetragnathidae						
Metellina sp.	-/2		-/4		-/5	
Metellina mengei	2/-	X	1/-	X		
Tetragnatha montana					1/-	V
Tetragnatha sp.	-/8		-/9		-/7	
Theridiidae						
Enoplognatha ovata	(9%) 6/19	V, VI	(9%) 7/10	V–VII	(8%) 3/8	V–VII
Robertus scoticus	1/-	IV				
Theridiidae un.			-/1			
Theridion sp.	-/7		-/12		-/7	
Theridion varians					1/-	VII
Thomisidae						
Diaea dorsata	(14%) 1/36	IV–VII, X	(7%) 1/12	IV–VII, XI	(4%) -/5	IV–VII
Ozyptila sp.					-/1	
Xysticus lanio			1/-	V		
Xysticus sp.	-/1		-/1			
Total no. of individuals	34/286		23/213		13/156	

Table 2. Observed species richness and species richness estimates for spider assemblages from tree branches in three developmental phases of oak–lime–hornbeam stand. Sampling completeness was calculated using Chao1 estimator.

	Optimal Phase	Terminal Phase	Regeneration Phase
Observed richness	17	16	11
Estimates			
Chao1 ± SD	41 ± 31	40 ± 31	19 ± 12
Chao2 ± SD	53 ± 44	45 ± 36	27 ± 21
Jackknife1 ± SD	25 ± 3	23 ± 2	16 ± 2
Jackknife2	32	29	20
Michaelis–Menten	22	20	15
Sampling completeness	41%	40%	58%

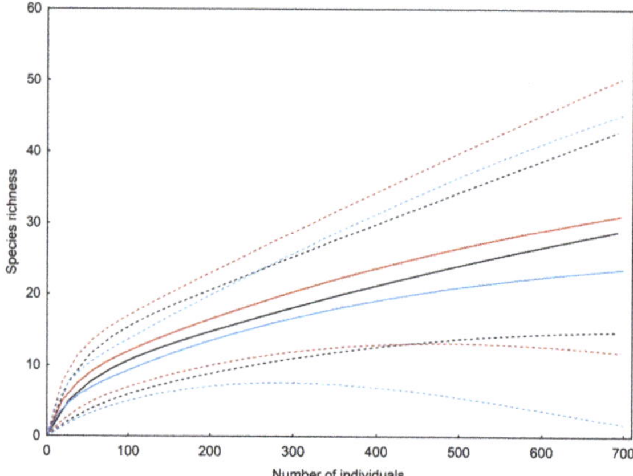

Figure 1. Individual-based rarefaction (solid) curves with 95% confidence limits (dashed curves) comparing species richness of branch-dwelling spider assemblages in three developmental phases of primeval oak–lime–hornbeam forest: the optimal phase (black), terminal phase (red) and regeneration phase (blue).

The highest species diversity of spider assemblages from tree branches was found in the terminal phase (H′ = 2.01), followed by the optimal phase (H′ = 1.91) and the regeneration phase of the forest stand development (H′ = 1.55). The Hutcheson test revealed differences between the optimal phase and the regeneration phase (t_{296} = 3.19; p = 0.002), as well as between the terminal phase and the regeneration phase (t_{281} = 4.10; $p < 0.001$), while no differences were found between the optimal phase and the terminal phase (t_{438} = 1.08; p = 0.283).

GLMs showed that the number of both adult and juvenile spider individuals, as well as the number of species (found in a given sample), was associated with the developmental phase of the forest stand and the sampling period (Table 3). Significantly more adult and juvenile spider individuals were found on the plot with the optimal phase compared to the regeneration phase (Figure 2). The number of adult individuals captured on branches decreased during the study period, i.e., from April to November (Figure 3a), in contrast to the number of juveniles, which increased significantly in the last two sampling months, i.e., October and November (Figure 3b). The number of spider species (found in a given sample) was significantly lower on the plot in the regeneration phase compared to the other plots (Figure 4a). The number of species was significantly higher at the beginning of the study period (April) compared to the other months (Figure 4b).

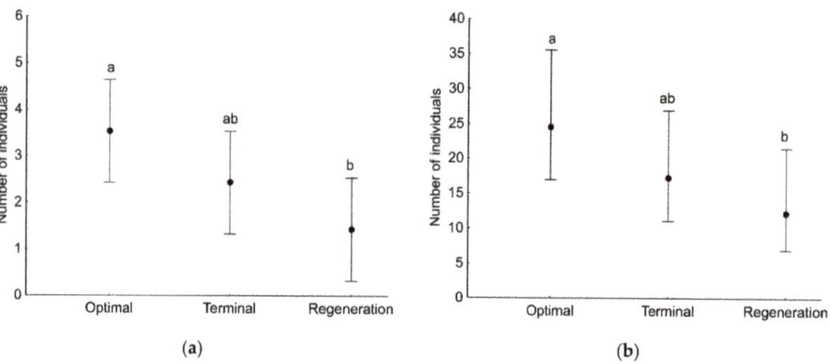

Figure 2. The number of adult (**a**) and juvenile (**b**) spider individuals (mean with 95% confidence limits) recorded on tree branches in a single sample in three developmental phases of primeval oak–lime–hornbeam forest. Different letters indicate significant differences between developmental phases.

Table 3. Results of generalised linear models assessing the effect of the developmental phase of the tree stand and sampling month on the abundance and the species richness of spider assemblages of tree branches.

Effect	Wald χ^2	df	*p*
Abundance of adult individuals			
Intercept	53.95	1	<0.001
Developmental phase	7.21	2	0.027
Sampling month	31.66	5	<0.001
Abundance of juvenile individuals			
Intercept	235.53	1	<0.001
Developmental phase	7.91	2	0.019
Sampling month	29.18	5	<0.001
Species richness			
Intercept	375.06	1	<0.001
Developmental phase	17.36	2	<0.001
Sampling month	16.01	5	0.007

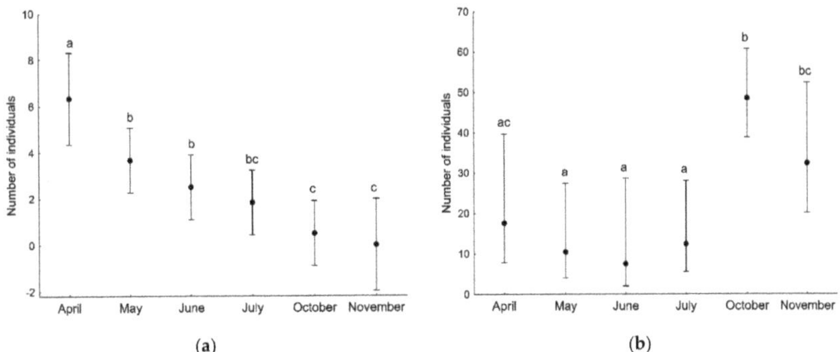

Figure 3. The number of adult (**a**) and juvenile (**b**) spider individuals (mean with 95% confidence limits) recorded on tree branches in particular sampling months. Different letters indicate significant differences between sampling months.

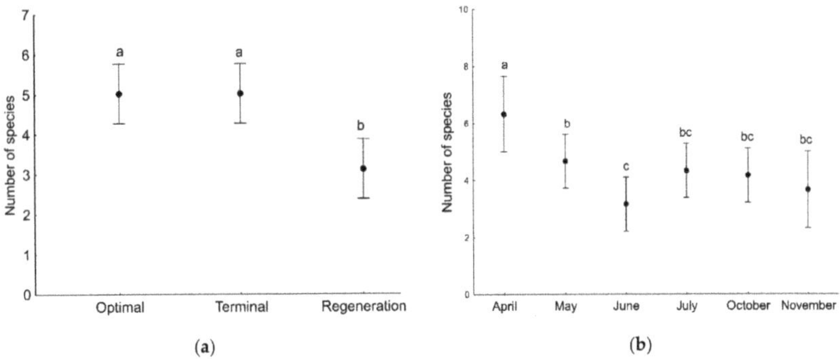

Figure 4. The number of spider species (mean with 95% confidence limits) recorded on tree branches in a single sample in three developmental phases of primeval oak–lime–hornbeam forest (**a**) and in particular sampling months (**b**). Different letters indicate significant differences between particular developmental phases (**a**) and particular sampling months (**b**).

3.2. Spiders of Tree Trunks

A total 2146 spider individuals belonging to 17 families were sampled on tree trunks during the study period (829 individuals in the optimal phase, 695 individuals in the terminal phase and 622 individuals in the regeneration phase). Juvenile spiders dominated in the collected material in each developmental phase (1845 individuals in total, ca. 86%). A total of 1610 individuals were identified to the species level: 621 from the optimal phase plot, 536 from the terminal phase plot and 453 from the regeneration phase plot. A total of 33 species were found (24 in the optimal phase, 23 in the terminal phase and 19 in the regeneration phase), of which 14 were common to all plots. A total of 13 species were represented by only 1 individual captured on a given plot (Table 4). The calculated estimators indicated higher species richness, especially for the terminal phase plot, where the sampling completeness was the lowest (Table 5). The most abundant spider species was *Anyphaena accentuata*, followed by *Amaurobius fenestralis*, in each phase of the stand development (Table 4).

Table 4. Spiders collected on tree trunks in three developmental phases of oak–lime–hornbeam stands in the Białowieża National Park (spider families, genus and species in alphabetical order). The percentages presented in parentheses, next to the number of individuals, show the proportion of each species. All individuals identified to the species level were included in the percentage composition, but values are only shown for species that reached at least 5%. Abbreviations: Ad./Juv.—number of adult/juvenile spider individuals, un.—individuals identified only to the family level. Roman letters indicate the months in which a given species was recorded.

Family/Genus/Species	Optimal Phase		Terminal Phase		Regeneration Phase	
	Ad./Juv.	Months	Ad./Juv.	Months	Ad./Juv.	Months
Agelenidae						
Agelenidae un.	-/1					
Coelotes atropos	5/-	VIII, IX	5/6	IV, VI–IX	2/-	VII, IX
Amaurobiidae						
Amaurobius fenestralis	(32%) 52/146	III–X	(31%) 62/106	III–X	(25%) 38/74	III–X
Anyphaenidae						
Anyphaena accentuata	(49%) 4/299	III–V, VII–X	(38%) 6/197	III–V, VII–X	(52%) 7/227	III–VI, VIII–X
Araneidae						
Araneidae un.					-/1	
Cyclosa conica	-/2	IV				
Nuctenea umbratica					2/-	VIII, IX
Clubionidae						
Clubiona caerulescens	1/-	V				
Clubiona lutescens	3/-	VII, IX	2/-	VII, VIII	4/-	VII, IX
Clubiona sp.	-/110		-/33		-/4	
Clubiona subsultans	1/-	X	3/-	X		
Dictynidae						
Dictyna sp.					-/1	
Gnaphosidae						
Haplodrassus cognatus	1/-	III	1/-	VI	1/-	V
Haplodrassus sp.	-/24		-/48		-/44	
Linyphiidae						
Agyneta ramosa			1/-	VI		
Drapetisca socialis	18/7	VI–X	11/13	VI–X	13/3	V–IX
Helophora insignis	1/-	VIII	1/-	VIII		
Labulla thoracica	1/5	VI–VIII	-/3	VII, VIII	1/-	IX
Lepthyphantes minutus	2/-	VIII, IX	7/1	VII–IX		
Lepthyphantes sp.	-/7					
Linyphiidae un.	-/15		-/22		-/39	
Lophomma punctatum	1/-	III				
Neriene clathrata					1/-	VI
Neriene montana	-/4	III, IV, VIII, X	-/15	IV, VIII–X	2/8	III, V, VI, VIII, X
Neriene sp.	-/2					
Savignia frontata			1/-	X		
Trematocephalus cristatus	-/11	III, IV, X			-/4	III
Lycosidae						
Piratula hygrophila					1/-	IX
Mimetidae						
Ero furcata			1/-	V		
Philodromidae						
Philodromus sp.	-/11		-/19		-/13	
Pisauridae						
Dolomedes fimbriatus	-/1	X				

Table 4. *Cont.*

Family/Genus/Species	Optimal Phase		Terminal Phase		Regeneration Phase	
	Ad./Juv.	Months	Ad./Juv.	Months	Ad./Juv.	Months
Salticidae						
Neon reticulatus			3/-	V, VII		
Neon sp.			-/1			
Salticidae un.			-/1		-/1	
Segestriidae						
Segestria senoculata	1/10	V, VIII–X	(9%) 5/43	V–X	(6%) 3/24	V–X
Tetragnathidae						
Metellina merianae			1/-	VII		
Tetragnatha sp.	-/2		-/1		-/2	
Theridiidae						
Dipoena nigroreticulata			1/-	VI	1/1	III, X
Enoplognatha ovata	2/1	VI	3/3	V–VIII	1/4	VI, IX
Steatoda bipunctata	1/2	VI, IX	2/11	V–IX	3/15	III, VII–X
Theridion mystaceum	1/-	V	3/-	V, VI	3/-	V–VII
Theridion sp.	-/28		-/34		-/58	
Platnickina tincta	1/2	VII, X	-/1	IV	2/1	VI, IX
Theridion varians	-/2	III				
Thomisidae						
Diaea dorsata	(5%) -/32	III, IV, X	-/18	III, IV, X	-/7	III, IV, X
Ozyptila praticola	1/-	VIII				
Ozyptila sp.	-/5				-/3	
Xysticus sp.	-/3				-/3	
Total no. of individuals	97/732		119/576		85/537	

Table 5. Observed species richness and species richness estimates for spider assemblages of tree trunks in three developmental phases of oak–lime–hornbeam stand. Sampling completeness was calculated using Chao1 estimator.

	Optimal Phase	Terminal Phase	Regeneration Phase
Observed richness	24	23	19
Estimates			
Chao1 $\pm$ SD	35 $\pm$ 10	55 $\pm$ 40	22 $\pm$ 3
Chao2 $\pm$ SD	43 $\pm$ 16	42 $\pm$ 19	21 $\pm$ 2
Jackknife1 $\pm$ SD	34 $\pm$ 4	32 $\pm$ 3	24 $\pm$ 2
Jackknife2	42	38	23
Michaelis–Menten	33	27	23
Sampling completeness	69%	42%	86%

The highest species diversity of spider assemblages from tree trunks was found in the terminal phase ($H' = 1.81$), followed by the regeneration phase ($H' = 1.58$) and the optimal phase of the forest stand development ($H' = 1.50$). The Hutcheson test revealed differences between the optimal phase and the terminal phase ($t_{1148} = 4.08$; $p < 0.001$), as well as between the terminal phase and the regeneration phase ($t_{947} = 2.85$; $p = 0.004$), while no differences were found between the optimal phase and the regeneration phase ($t_{985} = 0.94$; $p = 0.348$). The analysis of rarefaction curves revealed that the three developmental phases did not differ from each other in terms of the total species richness of spider assemblages inhabiting tree trunks (Figure 5).

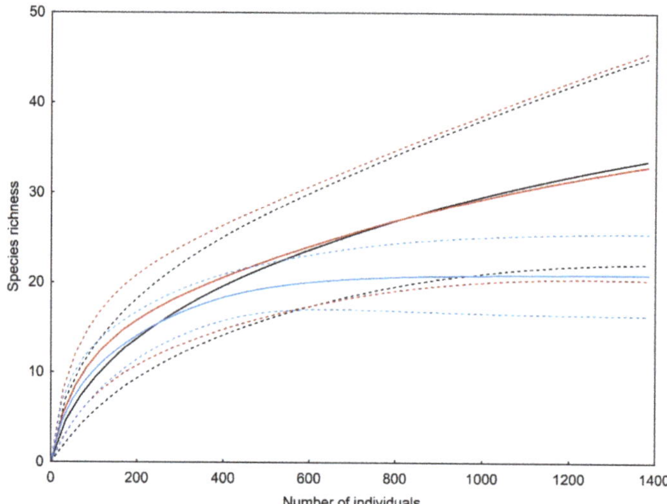

Figure 5. Individual-based rarefaction (solid) curves with 95% confidence limits (dashed curves) comparing species richness of trunk-dwelling spider assemblages in three developmental phases of primeval oak–lime–hornbeam forest: the optimal phase (black), terminal phase (red) and regeneration phase (blue).

The number of adult spider individuals and the number of species (found in each sample) were not associated with the developmental phase of the forest stand, while the number of juveniles was (Table 6). More juveniles were found in the optimal phase compared to the other two phases (Figure 6). Moreover, both the number of adults and juveniles were associated with the month of sampling, while number of species was not (Table 6). More adult individuals were captured in May and September (Figure 7a), while juveniles were significantly more numerous in March, April and October compared to the other months (Figure 7b).

Table 6. Results of generalised linear models assessing the effect of the developmental phase of the tree stand and sampling month on the abundance and the species richness of spider assemblages of tree trunks.

Effect	Wald χ^2	df	p
Abundance of adult individuals			
Intercept	183.62	1	<0.001
Developmental phase	3.84	2	0.146
Sampling month	138.80	7	<0.001
Abundance of juvenile individuals			
Intercept	583.29	1	<0.001
Developmental phase	6.81	2	0.033
Sampling month	51.06	7	<0.001
Species richness			
Intercept	362.97	1	<0.001
Developmental phase	2.14	2	0.342
Sampling month	4.11	7	0.767

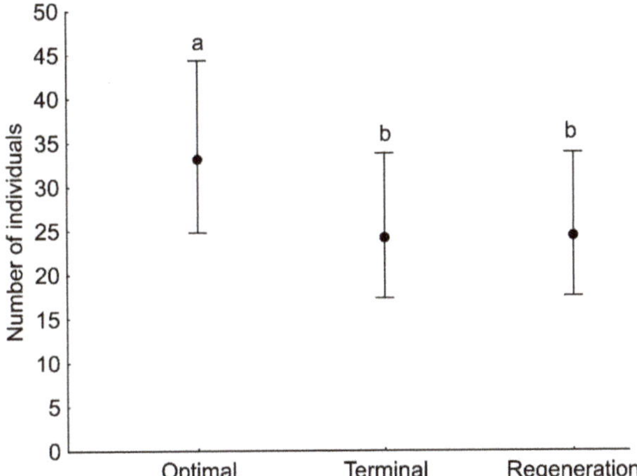

Figure 6. The number of juvenile spider individuals (mean with 95% confidence limits) recorded on tree trunks in a single sample in three developmental phases of primeval oak–lime–hornbeam forest. Different letters indicate significant differences between developmental phases.

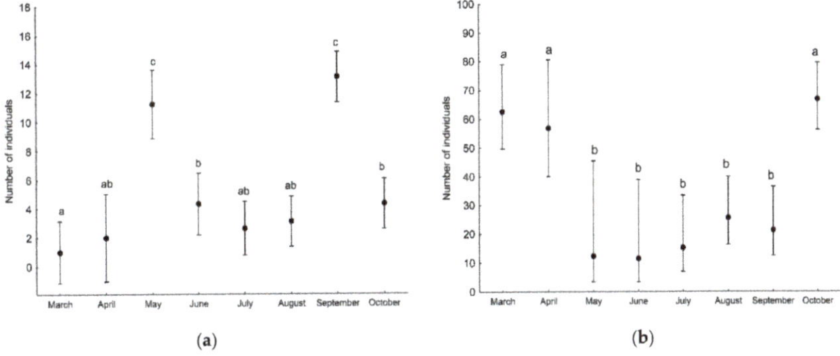

Figure 7. The number of adult (**a**) and juvenile (**b**) spider individuals (mean with 95% confidence limits) recorded on tree trunks in particular sampling months. Different letters indicate significant differences between sampling months.

4. Discussion

The hypothesis that the terminal phase, compared to the optimal and regeneration phases of stand development, would be characterised by higher spider abundance, species richness and species diversity was only confirmed for the last variable. We found the Shannon index was higher in the terminal phase compared to the other phases for both foliage-dwelling and trunk-dwelling spider assemblages. This fact may support our assumption that the most diverse niches for spiders exist in the terminal phase forest, as a result of significant disturbance in the stand structure. This disturbance was caused by the continuing process of old trees dying. As a result, some of the branches have broken off, and thus the crowns of the trees have become less dense. The structure of such a forest becomes very varied, with dead wood lying on the ground and lush herbaceous vegetation growing in places where gaps in the crowns have been created, and, at the same time, numerous trees of a large size provide varied niches for many groups of invertebrates. The phenomenon where significant variation in habitat structure translates into greater species diversity has been widely reported in the literature [25,35,36].

The present study showed that differences between the stand development phases do exist, but mainly indicated a lower number of spider individuals and number of species collected in individual samples on the plot in the regeneration phase compared to, above all, the plot in the optimal phase, where these variables reached the highest values. In our opinion, these differences may be explained, among other factors, by canopy cover, which was highest on the plot in the optimal phase and lower on the plot in the regeneration phase. The significant effect of canopy cover on spider assemblages in forests has been demonstrated by some authors. For example, Košulič et al. [20], studying epigeic spiders, showed that species richness was highest in places with medium canopy openness, while an open canopy supported the abundance of rare and threatened species. In addition, Oxbrough et al. [37] found that an open canopy favoured spider species typically absent in forest, and on a large scale, increased the abundance and species richness. In our study, we did not observe the occurrence of species associated with open habitats in the regeneration phase, but this may be due to the structure of the oak–lime–hornbeam forest in the BNP manifested by the fact that different developmental phases occupy relatively small areas located close to each other [11]. On the other hand, the canopy openness translated into the degree of herbaceous vegetation development in our study [18], and this may explain the differences between particular phases in abundance and species richness, at least for foliage-living spiders.

A higher number of juvenile and adult spiders in individual samples was collected on tree branches in the optimal phase compared to the regeneration phase, which may have resulted from poor herbaceous vegetation cover in the former phase [18]. The foliage can provide a kind of substitute for herbaceous vegetation, especially in the case of low-lying branches, and therefore, where herbaceous vegetation is less developed, spiders may be more likely to inhabit tree leaves. On the other hand, Stenchly et al. [38] found that the abundance and species richness of spiders from different strata, including those collected in tree crowns, were positively affected by herbaceous cover.

The sampled branches were located at a similar height from the ground as many herbaceous plants and to some extent resembled them in structure. This allowed us to assume that the fauna of spiders living on branches, at least those located not too high off the ground, should largely consist of plant-dwelling species. Our study largely confirmed this assumption. For example, *Trematocephalus cristatus*, the most abundant species on plots in the optimal and regeneration phases, was also abundant on herbaceous vegetation [18]. Other species, such as *Cyclosa conica*, *Enoplognatha ovata* and *Diaea dorsata*, were also collected both from tree branches and herbaceous vegetation in the study plots [18]. On the other hand, species such as *Linyphia triangularis* or *Bathyphantes nigrinus*, which were collected in large numbers on herbaceous vegetation, were not found on tree branches or only in small numbers. In addition, the fauna of foliage spiders should also include species living on tree trunks, as they can reach the leaves relatively easily. We found that *Anyphaena accentuata*, the most abundant spider on tree trunks, was also abundant on foliage, and was additionally collected on herbaceous vegetation [18]. However, the fact that more than half of the species found on branches were represented by only one individual on a given plot suggests that many species may have ended up there by chance and this is not their preferred habitat.

Only three species from the spider assemblages on tree trunks contributed 5% or more in each plot. Two of them, i.e., *Anyphaena accentuata* and *Amaurobius fenestralis*, clearly dominated, together accounting for 70–80% (dependent on plot) of all individuals identified to the species level. In addition, these species were collected in every (*Amaurobius fenestralis*) or almost every month (*Anyphaena accentuata*) of the trapping period. This showed that tree trunks are a common habitat for them, even though they may also live in others [5,18,39,40]. Furthermore, among the captured species, we can also include *Segestria senoculata* [40] and *Neriene montana* as typical inhabitants of tree trunks [23,41,42]. Other species were captured on tree trunks only in single months or in small numbers (more than 1/3 of the species were only represented on a given plot by a single captured individual). This shows that tree trunks serve as an incidental or temporary habitat for them, providing shelter or prey [43]. In addition, the fact that some species were recorded only in March, when winter still prevails in the Białowieża Forest, and/or in October when winter is approaching, may suggest that tree trunks are a wintering site for them. For example, *Diaea dorsata* was collected on tree trunks only in March, April and October, while it was found mainly on leaves from April to July. This may indicate that tree trunks are a substitute habitat for this species in the period when leaves have not yet appeared on trees and herbaceous vegetation has not fully developed, or that this is its overwintering site.

Spiders of tree crowns in the Białowieża Forest, outside primeval stands, were studied by Otto and Floren [7], who applied insecticidal knockdown fogging. They found the most abundant species were *Diaea dorsata* (21.8%), *Anyphaena accentuata* (16.1%), *Enoplognatha ovata* (13.5%) and *Paidiscura pallens* (9.9%). Of these species, we also found the first three in significant numbers on branches, although their proportions were different, while only *Anyphaena accentuata* was abundant on tree trunks. On the other hand, Otto and Floren [7] did not find *Amaurobius fenestralis* at all, which we found in large numbers on tree trunks. This may suggest that the fauna of spiders in tree crowns differs from that in tree trunks, whereas the fauna of spiders collected on branches is similar regardless of the height of the branches above the ground.

We found that the number of individuals, both adults and juveniles, and spider species (only in the case of foliage-living spiders) varied between the months evaluated in our study. This is certainly due to the phenology of individual spider species and changes in some habitat parameters (e.g., humidity, temperature) during the sampling period, although these were unfortunately not measured throughout the study period.

5. Conclusions

The analysed developmental phases of the oak–lime–hornbeam stand did not differ in terms of the total spider species richness. However, we found that species diversity of both foliage-dwelling and trunk-dwelling spider assemblages was higher in the terminal phase compared to other phases, which may indicate that this phase offers the most diverse niches for spiders due to the significant disturbance in the forest stand structure. The fauna of spiders inhabiting tree branches consisted largely of plant-dwelling species. We found that the fauna of tree trunks on each plot was dominated by two species—*Anyphaena accentuata* and *Amaurobius fenestralis*. For most spider species, tree trunks and branches are only temporary habitats or places where they can hide or overwinter.

Author Contributions: Conceptualisation, M.S. and T.S.; methodology, M.S. and T.S.; validation, M.S. and T.S.; formal analysis, T.S.; investigation, M.S.; resources, M.S.; data curation, T.S.; writing—original draft preparation, M.S. and T.S.; writing—review and editing, M.S. and T.S.; visualisation, T.S.; supervision, M.S. and T.S. All authors have read and agreed to the published version of the manuscript.

Funding: This research was funded by Siedlce University of Natural Sciences and Humanities (grant no. 76/20/B). The APC was funded by Siedlce University of Natural Sciences and Humanities.

Data Availability Statement: The data presented in this study are available on reasonable request from the corresponding author.

Acknowledgments: We would like to thank the authorities of the Białowieża National Park for their help during our study.

Conflicts of Interest: The authors declare no conflict of interest.

References

1. Nicolai, V. The bark of trees: Thermal properties, microclimate and fauna. *Oecologia* **1986**, *69*, 148–160. [CrossRef] [PubMed]
2. Koponen, S.; Rinne, V.; Clayhills, T. Arthropods on oak branches in SW Finland, collected by a new trap type. *Entomol. Fenn.* **1997**, *8*, 177–183. [CrossRef]
3. Floren, A. Abundance and ordinal composition of arboreal arthropod communities of various trees in old primary and managed forests. In *Canopy Arthropod Research in Europe*, 1st ed.; Floren, A., Schmidl, J., Eds.; Bioform Entomology: Nuremberg, Germany, 2008; pp. 279–298.

4. Sebek, P.; Vodka, S.; Bogusch, P.; Pech, P.; Tropek, R.; Weiss, M.; Zimova, K.; Cizek, L. Open-grown trees as key habitats for arthropods in temperate woodlands: The diversity, composition, and conservation value of associated communities. *For. Ecol. Manag.* **2016**, *380*, 172–181. [CrossRef]

5. Blick, T. Abundant and rare spiders on tree trunks in German forests (Arachnida, Araneae). *Arachnol. Mitt.* **2011**, *40*, 5–14. [CrossRef]

6. Stańska, M.; Stański, T.; Bartos, M. Spider assemblages of tree branches in managed and primeval deciduous stand of the Białowieża Forest. *Forests* **2022**, *13*, 5. [CrossRef]

7. Otto, S.; Floren, A. The spider fauna (Araneae) of tree canopies in the Białowieża Forest. *Fragm. Faun.* **2007**, *50*, 57–70. [CrossRef]

8. Mupepele, A.-C.; Müller, T.; Dittrich, M.; Floren, A. Are Temperate Canopy Spiders Tree-Species Specific? *PLoS ONE* **2014**, *9*, e86571. [CrossRef]

9. Tomiałojć, L. Characteristics of old growth in the Białowieża Forest, Poland. *Nat. Area. J.* **1991**, *11*, 7–18.

10. Tomiałojc', L.; Wesołowski, T. Diversity of the Białowieża Forest avifauna in space and time. *J. Ornith.* **2004**, *145*, 8–92.

11. Bobiec, A.; Van der Burgt, H.; Meijer, K.; Zuyderduyn, C.; Haga, J.; Vlaanderen, B. Rich deciduous forests in Białowieża as a dynamic mosaic of developmental phases: Premises for nature conservation and management. *For. Ecol. Manag.* **2000**, *130*, 159–175. [CrossRef]

12. Miścicki, S. Dynamics of the natural development phases of stands in the Białowieski National Park. *Sylwan* **2012**, *156*, 616–626.

13. Miścicki, S. Using developmentl stages of forest stands as a basis for forest inventory. *Sylwan* **1994**, *4*, 29–39.

14. Oxbrough, A.G.; Gittings, T.; O'Halloran, J.; Giller, P.S.; Smith, G.F. Structural indicators of spider communities across the forest plantation cycle. *For. Ecol. Manag.* **2005**, *212*, 171–183. [CrossRef]

15. Mullen, K.; O'Halloran, J.; Breen, J.; Giller, P.; Pithon, J.; Helly, T. Distribution and composition of carabid beetle (Coleoptera, Carabidae) communities across the plantation forest cycle—Implications for managements. *For. Ecol. Manag.* **2008**, *256*, 624–632.

16. Purchart, L.; Tuf, I.H.; Hula, V.; Suchomel, J. Arthropod assemblages in Norway Spruce monocultures during a forest cycle—A multi-taxa approach. *For. Ecol. Manag.* **2013**, *306*, 42–51. [CrossRef]

17. Trojan, P.; Bańkowska, R.; Chudzicka, E.; Pilipiuk, I.; Skibińska, E.; Sterzyńska, M.; Wytwer, J. Secondary succession of fauna in the pine forests of Puszcza Białowieska. *Fragm. Faun.* **1994**, *37*, 1–104. [CrossRef]

18. Stańska, M.; Stański, T. Plant-dwelling spider communities of three developmental phases in primeval oak-lime-hornbeam forest in the Białowieża National Park, Poland. *Eur. Zool. J.* **2021**, *88*, 706–717. [CrossRef]

19. Samu, F.; Lengyel, G.; Szita, É.; Bidló, A.; Ódor, P. The effect of forest stand characteristics on spider diversity and species composition in deciduous-coniferous mixed forests. *J. Arachnol.* **2014**, *42*, 135–141. [CrossRef]

20. Košulič, O.; Michalko, R.; Hula, V. Impact of canopy openness on spider communities: Implications for conservation management of formerly coppiced oak forests. *PLoS ONE* **2016**, *11*, e0148585. [CrossRef]

21. Stańska, M.; Stański, T.; Gładzka, A.; Bartos, M. Spider assemblages of hummocks and hollows in a primeval alder carr in the Białowieża National Park—Effect of vegetation structure and soil humidity. *Pol. J. Ecol.* **2016**, *64*, 564–577. [CrossRef]

22. Esquivel-Gómez, L.; Abdala-Roberts, L.; Pinkus-Rendón, M.; Parra-Tabla, V. Effects of tree species diversity on a community of weaver spiders in a tropical forest plantation. *Biotropica* **2017**, *49*, 63–70. [CrossRef]

23. Stańska, M.; Stański, T.; Hawryluk, J. Spider assemblages on tree trunks in primeval deciduous forests of the Białowieża National Park in eastern Poland. *Ent. Fenn.* **2018**, *29*, 75–85. [CrossRef]

24. Gallé, R.; Gallé-Szpisjak, N.; Zsigmond, A.-R.; Könczey, B.; Urák, I. Tree species and microhabitat affect forest bog spider fauna. *Eur. J. For. Res.* **2021**, *140*, 691–702. [CrossRef]

25. Malumbres-Olarte, J.; Vink, C.J.; Ross, J.G.; Cruickshank, R.H.; Paterson, A.M. The role of habitat complexity on spider communities in native alpine grasslands of New Zealand. *Insect Conserv. Diver.* **2013**, *6*, 124–134. [CrossRef]

26. Pierre, J.I., St.; Kovalenko, K.E. Effect of habitat complexity attributes on species richness. *Ecosphere* **2014**, *5*, 22. [CrossRef]

27. Áviala, A.C.; Stenert, C.; Rodrigues, E.N.L.; Maltchik, L. Habitat structure determines spider diversity in highland ponds. *Ecol. Res.* **2017**, *32*, 359–367. [CrossRef]

28. Hamřik, T.; Košulič, O. Impact of small-scale conservation management methods on spider assemblages in xeric grassland. *Agr. Ecosyst. Environ.* **2021**, *307*, 107225. [CrossRef]

29. Colwell, R.K. *EstimateS: Statistical Estimation of Species Richness and Shared Species from Samples*; Version 9.1.0.; University of Colorado: Boulder, CO, USA, 2019.

30. Chao, A.; Jost, L. Coverage-based rarefaction and extrapolation: Standardizing samples by completeness rather than size. *Ecology* **2012**, *93*, 2533–2547. [CrossRef]

31. Colwell, R.K.; Mao, C.X.; Chang, J. Interpolating, extrapolating, and comparing incidence-based species accumulation curves. *Ecology* **2004**, *85*, 2717–2727. [CrossRef]

32. Colwell, R.K.; Chao, A.; Gotelli, N.J.; Lin, S.-Y.; Mao, C.X.; Chazdon, R.L.; Longino, J.T. Models and estimators linking individual based and sample-based rarefaction, extrapolation and comparison of assemblages. *J. Plant Ecol.* **2012**, *5*, 3–21. [CrossRef]

33. Shannon, C.E.; Weaver, W. *The Mathematical Theory of Communication*; University of Illinois Press: Urbana, IL, USA, 1998; p. 144.

34. Hutcheson, T.P. A test for comparing diversities based on the Shannon formula. *J. Theor. Biol.* **1970**, *29*, 151–154. [CrossRef]

35. Hauser, A.; Attrill, M.; Cotton, P.A. Effects of habitat complexity on the diversity and abundance of macrofauna colonising artificial kelp holdfasts. *Mar. Ecol. Prog. Ser.* **2006**, *325*, 93–100. [CrossRef]

36. Lengyel, S.; Déri, E.; Magura, T. Species richness responses to structural or compositional habitat diversity between and within grassland patches: A multi-taxon approach. *PLoS ONE* **2016**, *11*, e0149662. [CrossRef]
37. Oxbrough, A.G.; Gittings, T.; O'Halloran, J.; Giller, P.S.; Kelly, T.C. The influence of open space on ground-dwelling spider assemblages within plantation forests. *For. Ecol. Manag.* **2006**, *237*, 404–417. [CrossRef]
38. Stenchly, K.; Clough, Y.; Tscharntke, T. Spider species richness in cocoa agroforestry systems, comparing vertical strata, local management and distance to forest. *Agric. Ecosyst. Environ.* **2012**, *149*, 189–194. [CrossRef]
39. Wunderlich, J. Mitteleuropäische Spinnen (Araneae) der Baumrinde. Zeitschrift für angewandte. *Entomologie* **1982**, *94*, 9–21.
40. Horváth, R.; Szinetár, C. Study of the bark-dwelling spiders (Araneae) on black pine (*Pinus nigra*) I. *Misc. Zool. Hung.* **1998**, *12*, 77–83.
41. Koponen, S. Spiders from groves in the southwestern archipelago of Finland (Araneae). *Rev. Ibérica Aracnol.* **2008**, *15*, 97–104.
42. Machač, O.; Tuf, I.H. Spiders and harvestmen on tree trunks obtained by three sampling methods. *Arachnol. Mitt.* **2016**, *51*, 67–72. [CrossRef]
43. Szinetár, C.; Horváth, R. A review of spiders on tree trunks in Europe (Araneae). *Acta Zool. Bulg.* **2005**, *Suppl. 1*, 221–257.

Article

Mechanics of the Prey Capture Technique of the South African Grassland Bolas Spider, *Cladomelea akermani*

Candido Diaz, Jr. [1,*] and John Roff [2]

[1] Biology Department, Vassar College, Poughkeepsie, NY 12604, USA
[2] Independent Researcher, Pietermaritzburg 3201, Kwazulu-Natal, South Africa
* Correspondence: cdiaz@vassar.edu

Simple Summary: Spiders are an order of organisms with highly diverse predatory techniques. All species produce silk and utilize varying degrees of adhesion to ensnare and trap prey long enough to envenomate them. Many spider families can be distinguished by their prey capture strategies, the silk structures they create, and the mechanical properties of the silk they spin. Understanding the diversity and function of these glues has much to teach us about natural bioadhesives and has application to our own synthetic adhesives. The most derived orb-webs are spun by bolas spiders, consisting of only a single capture thread, lined with a few glue droplets—often only one at the end. This web reduction must be accompanied by a strong glue. Additionally, the species *Cladomelea akermani* consistently spins its bolas and bounces. We use high-speed video to observe the prey-capture technique of *C. akermani*. The spider's willingness to spin allowed us to record and measure the kinematics of their unique bouncing, bolas spinning behavior, and overall prey capture technique. We then tested the additional hypothesis that this bouncing behavior serves an additional purpose in pheromone distribution by creating a computational fluid dynamics model. Spinning in an open environment creates turbulent air, spreading pheromones further and creating a pocket of pheromones. Conversely, spinning within a tree does little to affect the natural airflow.

Citation: Diaz, C., Jr.; Roff, J. Mechanics of the Prey Capture Technique of the South African Grassland Bolas Spider, *Cladomelea akermani*. *Insects* **2022**, *13*, 1118. https://doi.org/10.3390/insects13121118

Academic Editors: Thomas Hesselberg and Dumas Gálvez

Received: 20 October 2022
Accepted: 1 December 2022
Published: 5 December 2022

Abstract: Spiders use various combinations of silks, adhesives, and behaviors to ensnare prey. One common but difficult-to-catch prey is moths. They easily escape typical orb-webs because their bodies are covered in tiny sacrificial scales that flake off when in contact with the web's adhesives. This defense is defeated by spiders of the sub-family of Cyrtarachninae—moth-catching specialists who combine changes in orb-web structure, predatory behavior, and chemistry of the aggregate glue placed in those webs. The most extreme changes in web structure are shown by the bolas spiders which create only one or two glue droplets at the end of a single thread. They prey on male moths by releasing pheromones to draw them close. Here, we confirm the hypothesis that the spinning behavior of the spider is directly used to spin its glue droplets using a high-speed video camera to observe the captured behavior of the bolas spider *Cladomelea akermani* as it actively spins its body and bolas. We use the kinematics of the spider and bolas to begin to quantify and model the physical and mechanical properties of the bolas during prey capture. We then examine why this species chooses to spin its body, an energetically costly behavior, during prey capture. We test the hypothesis that spinning helps to spread pheromones by creating a computational fluid dynamics model of airflow within an open field and comparing it to that of airflow within a tree, a common environment for bolas spiders that do not spin. Spinning in an open environment creates turbulent air, spreading pheromones further and creating a pocket of pheromones. Conversely, spinning within a tree does little to affect the natural airflow.

Keywords: Cyrtarachninae; kinematics; arachnology; behavior; spin; environments

1. Introduction

Spiders are an order of organisms with highly diverse predatory techniques [1,2]. All species produce silk and utilize varying degrees of adhesion to ensnare and trap prey long enough to envenomate them [1]. Many spider families can be distinguished by their prey capture strategies, the silk structures they create, and the mechanical properties of the silk they spin [1,2]. Orb-weavers are largely a generalist group, known for their wagon wheel-shaped webs, which primarily focus on the capture of prey mid-flight [1,2]. The most derived spiders can produce upwards of seven different silks with unique mechanical properties; five of which are used in the construction of their webs [1,2].

As generalists, orb-weaving spiders typically eat any insect that is stuck in their web. One common but difficult-to-catch prey are moths. While plentiful in most ecosystems, the powder which covers their bodies consists of sacrificial scales which flake off when in contact with a spider's web [3,4]. As the moth thrashes, it releases them from the web, escaping. This niche has led to the evolution of a subfamily of spiders that have all evolved to specialize in moths only. The subfamily, Cyrtarachninae, consists of spiders that create various web structures sometimes limited to specific microenvironments [4–8]. Each of these species has adjusted their webs, behavior, and glue to overcome the superhydrophobic nature of a moth's body while reducing the overall structure and reliance on capture thread production [7–11]. Such examples consist of the genera *Cyrtarachne* and *Paraplectana* whose webs are horizontal with long dangling threads and *Pasilobus* whose webs are small, triangular, and only consist of three to four strands [6,7,9,11].

The most derived orb-webs are spun by bolas spiders, consisting of a single capture thread, lined with a few glue droplets—often only one at the end [12–16]. Bolas spiders consist of several genera all belonging to Cyrtarachninae [14,17]. They prey on male moths by releasing pheromones to draw them close, mimicking a female; they can even change the species of moth they are hunting throughout the evening [12,13,15,16]. From here, prey capture techniques seem to vary even between spiders using the same "weapons", such as the bolas wielders [18–21]. For instance, when a moth approaches the American species, *Mastophora hutchinsoni*, which holds a bolas consisting of one glue droplet very close to itself, it flicks the droplet at prey snagging it in a single strike [13,16,18,19]. These short-range bolas spiders respond only to the sound of their prey's wingbeats [22]. Other species are more active and less discerning. Species of the Genus' *Ordgarius* and *Exechocentrus* construct longer bolas of up to four droplets. Their attack behavior is easier to elicit, responding to the sound of human singing or a passing car, twirling their bolas in a circle [17,23–25]. The most active species, and the topic of this paper, belong to the genus *Cladomelea*. The South African grassland bolas spider species *Cladomelea akermani* constructs a bolas of up to three droplets and rotates its glue droplets and body during prey capture by spinning in a circular fashion. This behavior starts at sundown with or without prey [20,21,26,27]. While the unique behavior of many bolas spider prey capture systems has been observed in an ecological context, the exact kinematics of most bolas prey capture techniques have yet to be fully elucidated [13,15,16].

We are interested in the variation of the kinematics involved in prey capture, and how the behavior and mechanical properties of the glues utilized by moth-specialist spider species vary alongside it. In specific, we are interested in how prey capture techniques vary within bolas spiders species that construct varying bolas structures. Here, we use high-speed video to observe the kinematics of the prey-capture technique of *C. akermani*. Though we were unable to record a successful prey capture event, the spider's willingness to spin allowed us to record and measure the kinematics of their unique self-spinning/bouncing, bolas spinning behavior, and overall prey capture technique. We aim to understand the physical and mechanical differences in prey capture between this species and *M. hutchinsoni*, their non-spinning bolas-wielding American cousins. Firstly, we aim to verify the assumption that the spinning of the spider is directly related to the spinning of its bolas by measuring the speed, period, and rotational phases of the two. We then examine why *C. akermani* chooses to bounce during prey capture, an energetically costly behavior. Our hypothesis is that while the bouncing behavior may help to spin the glue droplet, it is also helpful to spread pheromones, widening its effective hunting area. We test this by creating a computational fluid dynamics model of airflow with and without a spinning spider within an open field and compared it to that of airflow in a tree environment, like that of *M. hutchinsoni* [15,16,18,19].

2. Materials and Methods

2.1. Field Observations and Kinematics of Prey Capture

Two adult female specimens of *C. akermani* were observed over three nights at Cumberland Nature Reserve (−29.513428, 30.505196, permit OP 2233/2022). Observations began at sundown as they transitioned from resting, to actively questing (waving their front legs in the air, Supplementary Video S1), to creating a bolas and spinning it. Once the spider had begun to make a bolas, a single Baslar acA1300–60gmNIR ACE camera was set up perpendicular to the horizontal plane of the spider; distances between the camera and the spider varied depending on the position of the spider relative to surrounding vegetation. Prey capture events were filmed at 83.6 frames per second (fps), the highest speed for the resolution of our camera, using a Fujinon 12.5 mm 2/3″ lens. Since most insects and arachnids do not rely on the red light for vision, subjects were illuminated using an ABI LED 54W near-infrared light (880 nm) to provide adequate lighting without impacting the behavior of the spider or moths [1].

C. akermani performs its capture technique without the direct presence of prey [21,27]. This allowed us to observe the creation and spinning of many bolas (N = 7) over only a few nights. From the videos, the movements of the spider and glue droplet were manually tracked using the open-source kinematics software Kinovea [28] (Figure 1). Up to four positions were tracked: (SA) the tip of the leg the spider uses to spin itself, (AA) the anchoring leg the spider holds the bolas with, and up to two glue droplets (D1, D2) located on the bolas. Each position was tracked during the entirety of each video which varied between 4 and 20 rotations in length. The position was tracked as a radial vector from a coordinate system located below the spider, perpendicular to the center of the spider. Rotational velocities were calculated using the finite difference in position and the known time interval between frames, measured as peak displacement using the measured diameter of rotation for that particular revolution. For videos in which the bolas were made of two droplets, we also measured the angle formed by the middle droplet, the end of the bolas, and the spider's anchoring leg using the law of cosines (Figure 1). For all values, averages and standard errors were calculated. A mixed ANOVA was run to determine if the four tracked locations varied statistically in their rotational period, velocity, and diameter. Radial distances were plotted and used to find the phase difference between the spider's legs and glue droplet maximum displacement.

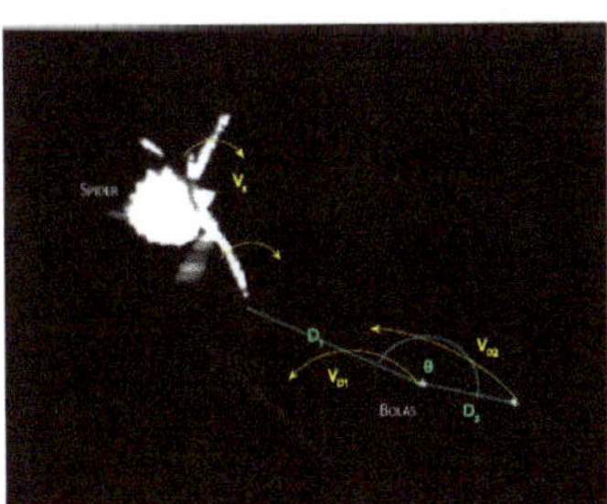

Figure 1. Tracking of Prey Capture Technique. Videos were tracked using Kinovea. Up to four locations were manually digitized for the position which was used to calculate the rotational velocity (V), period, and rotational diameter. The angle formed by the glue droplets was calculated over the course of the video using the law of cosines and the measured distances. The coordinate system is located below the spider perpendicular to the anchoring leg of the spider.

2.2. Pheromone Airflow—Computational Fluid Dynamics Model

When sundown begins, *C. akermani* sets up its bolas and begins to spin, continuing to do so with or without the apparent presence of prey. This raised a question of whether or not this bouncing behavior serves a purpose other than to spin their bolas, as this is a timely and energetically costly behavior that is not shared with every other bolas species [16,17,27].

We made simple models of the environments of *C. akermani* and *M. hutchinsoni*, an open grassland, and a large arboreal structure respectively, using Autodesk Fusion 360 (Figure 2A) [29]. These models were then imported as .step files into Autodesk CFD 2023 where four varying flow condition tests were performed: (1) a small breeze at 3 m/s, (2) slow diffusion at 0.05 m/s, (3) the magnus effect of a spinning spider at 3 m/s, and (4) the combination of a spinning spider and a breeze (details listed below) [30].

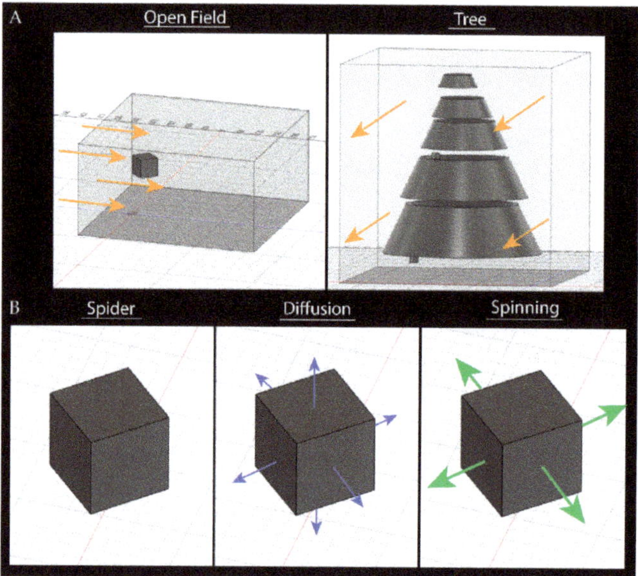

Figure 2. Environment models and flow conditions used in computational fluid dynamics modeling. (**A**) Open field tests involved a rectangular environment open on all sides except the bottom. Orange arrows show the direction of the wind, behind the spider, at 3 m/s. The tree model was made of five rings, and two spiders were placed due to being a larger structure, one at the top and one at boom. (**B**) The spider and pheromone cloud were modeled as a perfect cube, with sides of length 1 cm. The diffusion condition applied a small velocity, 0.05 m/s, exiting all faces of the cube. The spinning condition applied a larger velocity, 3 m/s, exiting the four front and back faces of the spider.

In each model, the spider was represented as a cube-shaped volume of pheromones, using the default material properties of 'Air' material (Figure 2B). The field model consisted of a large open rectangular area with only boundary conditions on the floor. Within the field, results were shown as the airflow velocity along three planes parallel to the wind: one in the plane of the spider, one slight above, and one slightly below. Our tree model is composed of four increasingly smaller plates, forming a cone shape, to mimic the open space of the leaves where the spider hangs. For the tree conditions, two spiders were placed within the confines of the tree with one at the bottom and one at the top (Figure 2A). The flow is shown by four planes (1) bisecting the tree, parallel to the tree trunk, (2) perpendicular to the wind flow, just before the wind contacts the tree, (3) a plane parallel to the flow of wind and located directly below the opening of the canopy, and (4) a plane parallel to the wind bisecting the tree halfway up.

Condition 1. Wind—We model air flow within the field as laminar air flow through a wind tunnel open on all sides, originating behind the spider (within the field) and/or tree, 3 m/s (Figure 2A).

Condition 2. Diffusion—We model the diffusion as a very low distribution over time in all directions, 0.05 m/s (Figure 2B). Within our models, this diffusion was modeled as originating out of the spider in all directions, as 6 sides of a cube. For this behavior, we expect the flow to be extremely low and slow. Diffusion can be described by Fick's first law which relates the diffusive flux to the gradient of the concentration, in that a solute will move from a region of high concentration to a region of low concentration across a concentration gradient [31]. These low forces cause diffusion to be a relatively slow method of distributing pheromones. Equation (1) shows the general form of diffusion flux, J, the amount of substance per unit area per unit time. D is the diffusion coefficient, φ is the concentration, and x is the position.

$$J = -D\frac{d\varphi}{dx} \tag{1}$$

Condition 3. Bouncing—As *C. akermani* bounces its large legs, it creates areas of low pressure which forces the pheromones around to move away from the spider. This effect, similar to a rotating baseball, except that now occurs in both directions as the spider alternates, is known as the Magnus effect [32]. The air flow speed is shown by Equation (2) where s is the rotation rate (revolutions per second), ω is the angular velocity of spin (radians/second) and r is the radius of the cylinder (meters). This means the resulting magnus effect is proportional to the spinning speed and the airflow behind them.

$$G = (2\pi r)^2 * s = 2\pi r^2 \omega \tag{2}$$

This equation shows that the airflow created by the spider should be outwards from its motion, like that coming off a hand fan. To model this cyclical bouncing, we used air flow velocity pointing outward and parallel to the ground. We model this as vectors flowing outward of the four faces in the plane of airflow. This excludes the faces pointing to y-axis. As the spider moves, it creates a wafting force in all directions, half within the direction of the wind and half against it. Because it is bouncing and not spinning the force alternates left and right, fanning the pheromones, instead of consistently pushing it in one direction, 3 m/s (Figure 2B).

Condition 4. Bouncing and Wind—This condition is the most equivalent to that of the *C. akermani* and utilizes both wind and bouncing conditions listed above at the same time.

3. Results

3.1. Observations on Bolas Building Behavior and Kinematics of Prey Capture

Cladomelea akermani sits on a grass stalk or leaf until sundown when it begins to be active, waving its long legs into the air before moving from its resting spot. It starts creating a bolas by first moving up and along the top of its leaf or grass stalk and creating a simple beam. It then reinforces the beam, moving across it, as all other spiders do during web preparation [1]. As it does this, it waves its long legs in the air, wafting them about (Supplementary Video S1). This could be for multiple reasons, which are untested, including sensing for moth activity on the breeze, testing the wind speed, and/or distribution of pheromones. After the beam is reinforced *C. akermani* moves towards one side of the beam, releasing silk. At a distance of ~5–6 cm, the spider lowers itself, extruding another thread. There it dangles its silk, now connected to two threads, creating a V-shape. Dropping a bit further the spider begins to attach a bolas to the strand it is not dangling from. Pulling out large amounts of flagelliform and aggregate glue, the spider pushes it together into a large ball, letting the bolas slowly fall away as it gets larger, ultimately swinging away (Figure 3A, Supplementary Video S2). When the spider decides to make a bolas with multiple glue droplets, it then moves back up to the top anchoring an additional thread and making another v-formation (Figure 3B, Supplementary Video S3). It lowers itself again but not as far as previously. It repeats its glue droplet-making process and the droplet is allowed to swing into the previously constructed strand, forming a single overall structure (Figure 3C).

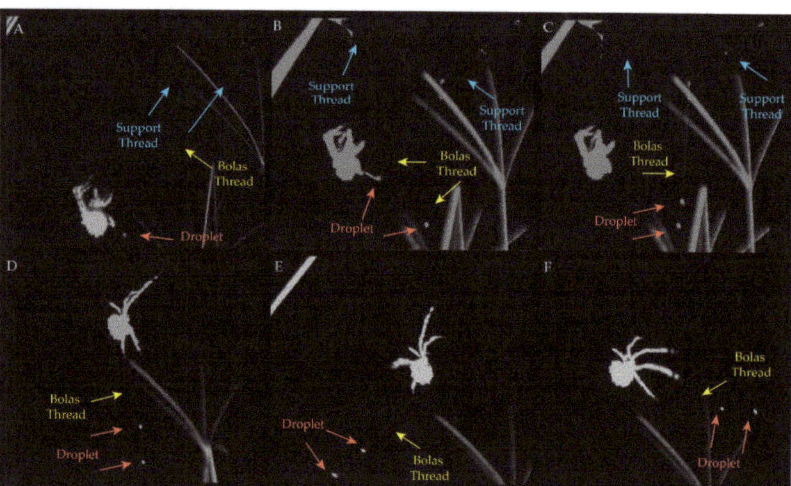

Figure 3. Breakdown of Multi-Bolas Creation and Prey Capture Behavior. The spider shown has a carapace width of 12 mm. Red arrows in each image are glue droplets and yellow arrows show the bolas capture thread. *C. akermani* sits on its grass or leaf until sundown. Like other spiders it first moves up and along the top of its leaf or grass stalk, creating a simple beam. It then waves its long legs in the air, wafting them about. (**A**) the spider creates two support threads (blue arrows), sliding down one and attaching a bolas to one strand. It lets the droplet slowly fall away as it gets larger. (**B,C**) then moves back up to the top anchoring another thread to the support beam. It drops back down again, but not as far as previous, and creates another glue droplet, which is allowed to swing into the other strand, making a single overall structure. (**D**) Once finished, the spider orients itself parallel to the support bar, grabs the bolas with its lower second leg, and holds its especially long and hairy front legs out. (**E**) It spins the bolas in a circle with its anchoring second leg. (**F**) after it has a consistent and circular motion started with the droplets, the spider begins to swing its front legs back and forth, spinning/bouncing.

Once finished, the spider orients itself parallel to the support bar, grabs the bolas with its lower second leg, and holds its especially long and hairy front legs out (Figure 3D). It spins the bolas in a circle with its anchoring leg, second leg (Figure 3E). After it has a consistent and circular motion started with the droplets, the spider begins to swing its front legs back and forth, bouncing on the line, twisting but never fully rotating (Figure 3F, Supplementary Videos S4 and S5).

We were not able to record a moth actively being caught with our camera during this trip, but we did observe moths' approach. They arrived rapidly from several meters downwind of the spider and flew in an indirect almost spiral fashion, towards the spider. Though close the spinning bolas failed to strike the moth, which then flew away. Throughout the evenings, very few moths closely approached the spider, though there was observable moth activity in the area immediately surrounding the spider all night. Moths were observed doing numerous things, such as sitting in the grass directly below the spider and surrounding it. Some moths climbed the grass stalks the spider was anchored on, moving up and down them, while fluttering their wings. However, those moths never approached the spider or flew near it.

Throughout our seven recorded videos, all measured values (period T, velocity v, and radius r) were highly variable for both the spider and bolas glue droplets (Figure 4). Post hoc Tukey analysis showed the anchoring leg (AA) was always statistically lower in all values than that of the swing leg (SA) of the spider and the glue droplets. Within each video, the maximum rotation of D_2 was always greater than D_1, 1.28 ± 0.13 times. The velocity similarly was always 1.37 ± 0.11 times higher, while the period was the same at 1 ± 0.21 s. D_2 had the overall largest values with a maximum rotational diameter of 11.52 cm, and spinning upwards of 150.09 cm/s. The angle formed by D_1, D_2, and AA showed a variation between the 180° of a straight line and 39.9°.

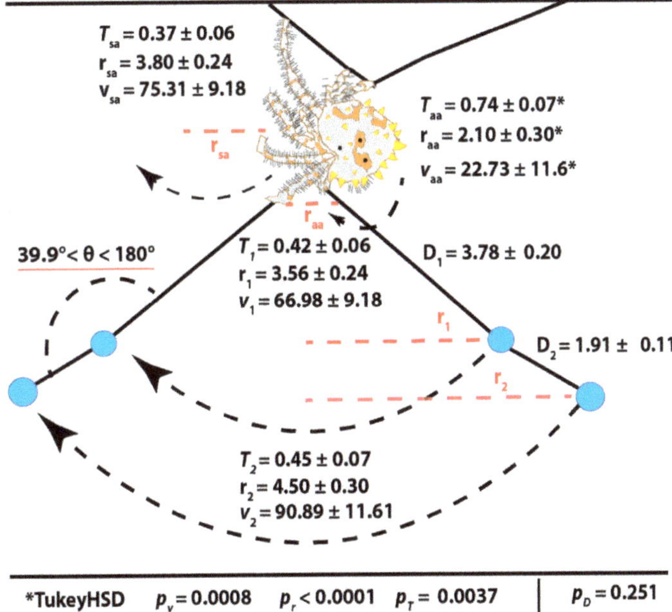

Figure 4. Average Kinematics of Spider Legs and Bolas Glue Droplets During Spinning. *Post hoc* Tukey analysis showed that the spinning of the anchor leg is the most statistically different for all measured values (period T in s, velocity v in cm/s, and radius r in cm). The anchor leg barely moves and completes only half a rotation for each one of the glue droplets that the swing leg completes. Though the large variation in spinning speed led to no overall difference within tests, the D_2 was always spinning faster and further out than D_1. The angle between D_1 and D_2 can be seen changing while spinning leading to variation in the angle between them (Supplementary Data File S1, Excel Datasheet).

The swing leg bounces only after a firm rotation of the glue droplets has been achieved by slowly spinning the anchor leg. Then the spider begins to throw its body, increasing glue droplet rotational diameter and speed, confirming the swing leg is aiding but not necessary to rotate. Once the glue droplets are in motion and the spider begins to swing its body. This syncs the rotational period for the droplets and swing leg (Figure 5). This means the spider is thrusting its body forward and then bouncing back in the time it takes for a single rotation of the glue droplets. Calculating the phase lag between maximum displacements showed the highest displacement of the swing leg always occurred one or two frames before the maximum of each glue droplet. The period of the anchor leg was twice of others, meaning it is not fully in sync and not the driving force behind of rotation during swinging (Figure 5).

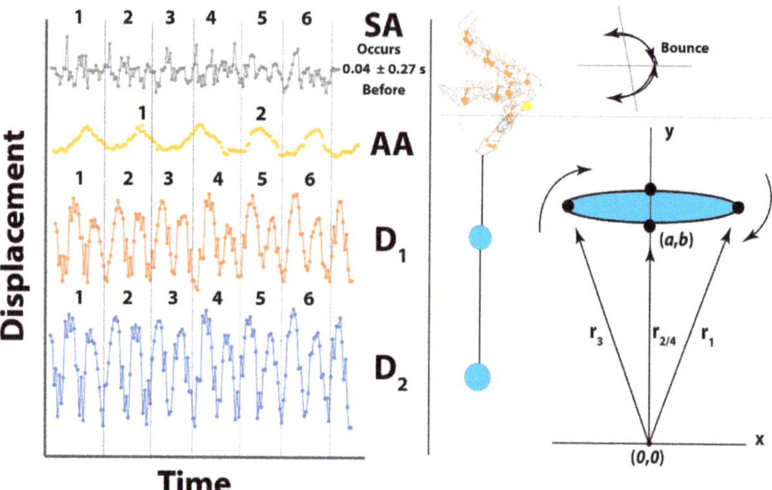

Figure 5. Leg and Glue Droplet Rotational Phase Alignment. Because all distances were measured as vectors radiating from a common origin, rotational periods are shown as two maxima, as the maximum displacement occurs twice. The second rotation or the second half of the circle is always shorter and slower, meaning that the energy the spider is putting into the thread is falling off almost instantly. Thus, two peaks show the period, and the rotational diameter is the maximum displacement of both peaks combined. Calculating the frames between distances, we found that the displacement of droplet two (D_2) was always higher than droplet one (D_1) but D_1, D_2, and the swing leg (SA) were nearly perfectly in sync. The maximum displacement of the SA always occurred two frames or 0.04 s before the droplets, meaning it was using its body to accelerate the glue droplets. The SA does not have two peaks because the spider is not completing a full rotation, but instead is thrusting forward and bouncing back. The period of the anchor leg (AA) was always twice as long as the other periods.

3.2. Pheromone Airflow—Computational Fluid Dynamics Model

Diffusion creates minimal velocity around the spider, most easily visualized in open field conditions (Figure 6). The spider's bouncing and spinning creates a larger velocity in front, and a lower velocity behind it, creating turbulence. This forces pheromones to distribute further above, below, and in front of the spider. The largest airflow was created by the wind in both test environments. In the field, the wind will slow down slightly as it passes through the nearly stagnant pheromone cloud, but the streams remain laminar and independent (Figure 6). When spinning is added to the wind, the velocity in front of the spider is increased from 300 cm/s to 350 cm/s. The normally straight pheromone trail becomes turbulent, as air downwind becomes mixed both above and below the plane of the spider. These trails are further carried downwind, now in multiple directions. The spinning additionally forces pheromones behind the spider to collide with the oncoming wind. This forces air upward behind the spider at 400 cm/s. This would further distribute the pheromones over the open landscape. The various barriers and openings within the tree innately create pockets of higher, 500 cm/s, and lower-density air as the wind blows past it (Figure 6). The wind is pulled through the tree, swirling around, before being pulled out the opposite side. The highest flow velocities are surrounding the tree and exiting the tree downwind. The velocities within the tree fall when the spider bounces, slowing down the air flow in the tree to 250 cm/s.

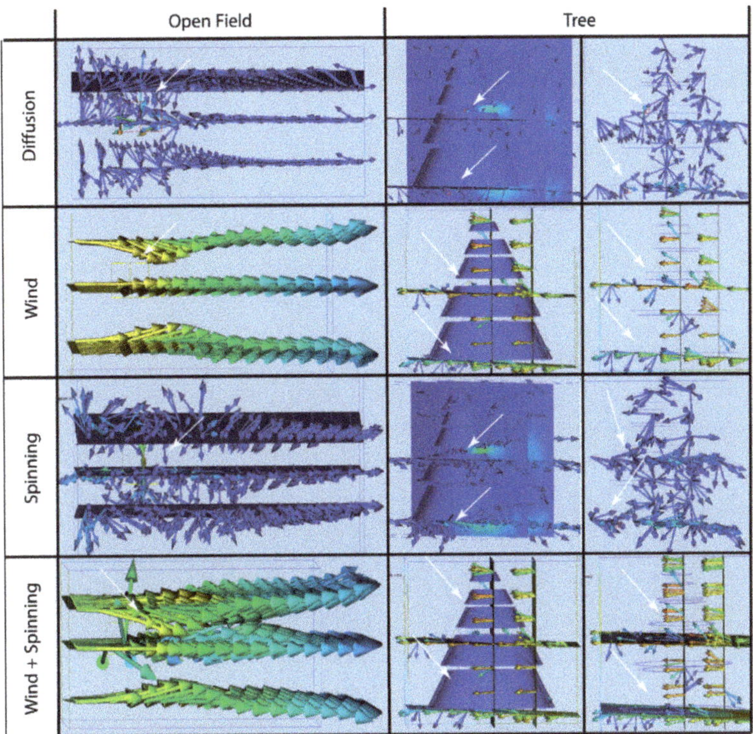

Figure 6. Pheromone Airflow—velocity field results of open field and tree conditions. Each image shows the Autodesk CFD velocity vector results. Air velocities are coded by a color gradient where blue is slower, and red is faster (colors are relative within each test). White arrows show the location of the spider in each test, one in the field and two in the tree. The tree tests are shown with the full tree structure and with only an outline of the tree to view airflow within the tree more easily. Open field conditions—The open field shows how diffusion creates minimal velocity around the spider. The spinning creates a larger velocity but mostly creates turbulence around the spider which continues further downstream. The wind has the largest effect on airflow. It slows down slightly as it passes through the nearly stagnant pheromone cloud, but the stream remains laminar and independent. When spinning is added to the wind, the velocity in front of the spider is increased. The normally straight trail of pheromones is instead mixed up and down being carried by the wind in now multiple directions, further distributing the pheromones over the open landscape. Tree conditions—The various barriers and openings within the tree innately create pockets of higher and lower-density air as the wind blows past it. The wind is pulled through the tree, swirling around, before being pulled out the opposite side. The highest flow velocities are surrounding the tree and exiting the tree downwind. The structure of a tree allows pheromones to be distributed throughout a tree more easily and naturally. Pheromones fill the tree and then leak from it, creating a beacon. The velocities within the tree fall when the spider bounces, slowing down the air flow in the tree.

4. Discussion

To our knowledge, we perform here the first kinematic descriptions of the spinning behavior of the bolas spider *Cladomelea akermani*. We used a high-speed video camera to observe their prey capture technique and track the spider's legs as it actively bounced its body and spun its bolas. Through this, we confirmed the previously stated hypothesis that the bouncing behavior of the spider helps to increase the spinning speed and rotational diameter of the bolas; observed as the largest displacement of the swing legs always occurring before the droplet's highest velocity and displacement. We then tested the additional hypothesis that this spinning behavior, an energetically costly one, serves an additional purpose in pheromone distribution; we tested this by creating a computational fluid dynamics model of airflow within an open field and compared it to that of airflow within a tree, a common environment for bolas spiders that do not spin.

Our question was, what is advantageous about the spinning behavior of *C. akermani* and how does it correlate with their hunting environment? Why is this additional energy expenditure worth it? To answer this, we looked at the flow of pheromones in their respective environments. We believe differences in behavior correlated with two things (1) environment and (2) its resulting effect on the approach behavior of moth prey. At base, the structure of the environment influences a species' exposure to ambient conditions, such as temperature and airflow, and influences the distribution of species, both predator and prey. Relying on pheromones, bolas spiders are especially susceptible to changes in prey distribution and wind. The habitat and use of a bolas by the previously studied species *Mastophora hutchinsoni* both serve as a great minimalist comparison to *C. akermani*. For both species, the key first step in prey capture is attracting the moth to themselves.

A tree seems to be a simpler hunting environment for bolas spiders than an open field. This statement is based on two factors (1) natural airflow and (2) pheromone interaction with other bolas spiders. When *M. hutchinsoni* sits within a tree, its pheromones are contained within the canopy of the tree, except for the air which billows out from under the tree leaves (Figure 6). Our models predict that spinning within a tree does little to affect the natural airflow, as the tree structure directs airflow, making spider movement redundant. The presence of multiple spiders seems advantageous in this way, turning the tree into a pheromone beacon to the communal hunting ground. Once the moths are within the tree, they can follow the natural pheromone gradients to find the closest spider. They then hover close to the spider, sometimes even touching it [18,33,34]. Thus, it makes sense this species creates significantly shorter bolas, and with only a single glue droplet [5,15,16]. Only when the moth is near it, does the spider flick its bolas. This simplicity in hunting is a consequence of the moths' ease at finding the spider.

For *C. akermani*, its hunting habitat is much more complicated as it sits in an open field, especially susceptible to wind. Persistent and strong winds, blow its small and slowly diffusing pheromone cloud in multiple directions, thinning it out and complicating the following of its gradient. Changes in airflow direction and intensity can lead to varying and cluttered gradients, limiting the ability of the moth to locate the spider. This confusion could be especially apparent when multiple spiders are positioned close to one another. Wind can mix the pheromone trails of spiders too close to one another, perhaps sending prey in the wrong direction or towards another spider increasing the potential of intraspecies competition. From our models, the spinning of the spider fights these issues with the wind. As the spider spins, it creates flow in all directions surrounding itself. When in the direction of the wind, it helps spread pheromones forward but when fighting against the direct flow of wind pheromones are sent upward and downward while slowing down airflow in the immediate area (Figure 6). Resistance to flow from the spinning may keep pheromones from simply blowing away and making it easier to target for the moth. In this environment the spinning is doing double duty, spinning the bolas and controlling pheromone distribution.

The bolas created by *C. akermani* are three to five times longer than that created by *M. hutchinsoni* which may be a response to this inability of the moths to directly locate the spider. These bolas are generally made with two glue droplets but can sometimes be made of one or three [20,21,27]. The small distance of 2 cm between the droplets creates a larger effective zone of adhesion. Any contact with the space between the glue droplet will lead to the collapsing of the bolas around the prey, ensnaring it in both glue droplets. The spinning of the bolas takes these two linear cms of glue and rotates it about an axis. The spider spins the glue droplets twice a second and creates a larger apparent cone of adhesion. Bouncing allows the adhesive zone to spin faster and overall wider area. Thus, unfortunately, there is little observational evidence, aside from our own, to determine if captured moths are typically hovering near the spider, such as those caught by *Mastophora*, or are more aggressively flying—necessitating a rapidly moving bolas. A second advantageous consequence in the structure of the multi-droplet bolas, but potentially accidental, is the wobble and bending around the D_1 glue droplet. The glue droplets are constructed of viscoelastic glue and extra silk thread, filling a liquid droplet [35–37]. The construction of the bolas using two separate threads, creates a system of two tensions, adhered by a liquid droplet acting as a joint. The bending of the droplets leads to wobbling and when spun quickly can help to further increase the effective area of the adhesive strike zone.

Here, we attempt to correlate differences in bolas construction and prey capture behavior between *C. akermani* and *M. hutchinsoni* with differences in the hunting environment. We believe that alterations such as elongation of the bolas, adding additional droplets, and spinning are advantageous when living in a large open grassland where wind can readily but randomly disperse pheromones. Such adaptations are not necessary for species such as *M. hutchinsoni* which live more enclosed environments where moths may already gather, such as a tree. The alterations to the bolas of *C. akermani* are closely tied to the behavior of its prey and future studies should aim to record the behavior of a moth being captured, as this will inform our understanding of why bolases are spun. As of now, we cannot rule out several alternative hypotheses for adaptive advantages of spinning, such as the spinning behavior being tied to a unique size or physical property of its target prey species. It also will let us confirm our hypothesis that the rotation allows the glue droplets to roll around their prey when struck, wrapping them. Observing these species with multiple cameras would also allow us to calculate the forces placed on the silk and calculate the associated material properties. We would also like to collect and test these bolas for biomechanical and chemical analysis to compare the structure and diversity of the glues within the moth-specialist subfamily Cyrtarachninae. The material diversity within the family is shown by the glue droplets of *C. akermani* which are capable of remaining as distinct droplets on a single strand without flowing with gravity into one another. Similar to *M. hutchinsoni* these glues appear thicker than others such as *Cyrtarachne akirai*'s low viscosity glue which readily flows [4,18,38]. Understanding the diversity and function of these glues has much to teach us about natural bioadhesives and has application to our own synthetic adhesives.

Supplementary Materials: The following supporting information can be downloaded at: https://www.mdpi.com/article/10.3390/insects13121118/s1. Video S1: leg waving; Video S2: creating a single bolas; Video S3: creating a double bolas; Video S4: swinging a single bolas; Video S5: swinging a double bolas. Supplementary Data File S1: The excel data sheet used to create Figure 3 can be downloaded as Supplementary Material. Supplementary Data File S2: Fusion CFD files can be downloaded for the swing + bounce condition for the field and tree.

Author Contributions: Conceptualization, data curation, formal analysis, visualization, writing—original draft preparation, funding acquisition, C.D.J.; methodology, validation, investigation, writing—review and editing, C.D.J. and J.R. All authors have read and agreed to the published version of the manuscript.

Funding: This work was supported by the National Science Foundation award IOS-2031962. South Africa research permit #OP 2233/2022.

Data Availability Statement: The excel data sheet used to create Figure 3 can be downloaded as Supplementary Material.

Acknowledgments: We would like to thank the Cumberland Nature Reserve for allowing us to use their land to locate and film spiders. Candido would also like to thank Aubrey Clark for editing and for all her support and love.

Conflicts of Interest: The authors declare no conflict of interest.

References

1. Foelix, R.F. *Biology of Spiders*; Oxford University Press: New York, NY, USA, 2011.
2. Blackledge, T.A.; Scharff, N.; Coddington, J.A.; Szüts, T.; Wenzel, J.W.; Hayashi, C.Y.; Agnarsson, I. Reconstructing web evolution and spider diversification in the molecular era. *Proc. Natl. Acad. Sci. USA* **2009**, *106*, 5229–5234. [CrossRef] [PubMed]
3. Nentwig, W. Why do only certain insects escape from a spider's web? *Oecologia* **1982**, *53*, 412–417. [CrossRef] [PubMed]
4. Diaz, C.; Maksuta, D.; Amarpuri, G.; Tanikawa, A.; Miyashita, T.; Dhinojwala, A.; Blackledge, T.A. The moth specialist spider Cyrtarachne akirai uses prey scales to increase adhesion. *J. R. Soc. Interface* **2020**, *17*, 20190792. [CrossRef]
5. Levi, H.W. *The Bolas Spiders of the Genus Mastophora (Araneae: Araneidae)*; Harvard University: Cambridge, MA, USA, 2003.
6. Tanikawa, A. Two new species of the genus Cyrtarachne (Araneae: Araneidae) from Japan hitherto identified as *C. inaequalis*. *Acta Arachnol.* **2013**, *62*, 95–101. [CrossRef]
7. Baba, Y.G.; Kusahara, M.; Maezono, Y.; Miyashita, T. Adjustment of web-building initiation to high humidity: A constraint by humidity-dependent thread stickiness in the spider Cyrtarachne. *Naturwissenschaften* **2014**, *101*, 587–593. [CrossRef]
8. Diaz, C.; Tanikawa, A.; Miyashita, T.; Amarpuri, G.; Jain, D.; Dhinojwala, A.; Blackledge, T.A. Supersaturation with water explains the unusual adhesion of aggregate glue in the webs of the moth-specialist spider, Cyrtarachne akirai. *R. Soc. Open Sci.* **2018**, *5*, 181296. [CrossRef]
9. Diaz, C.; Tanikawa, A.; Miyashita, T.; Dhinojwala, A.; Blackledge, T.A. Silk structure rather than tensile mechanics explains web performance in the moth-specialized spider, Cyrtarachne. *J. Exp. Zool. A Ecol. Integr. Physiol.* **2018**, *329*, 120–129. [CrossRef] [PubMed]
10. Andersen, S.O. Biochemistry of insect cuticle. *Annu. Rev. Entomol.* **1979**, *24*, 29–59. [CrossRef]
11. Cartan, C.K.; Miyashita, T. Extraordinary web and silk properties of Cyrtarachne (Araneae, Araneidae): A possible link between orb-webs and bolas. *Biol. J. Linn. Soc.* **2000**, *71*, 219–235. [CrossRef]
12. Eberhard, W.G. Aggressive chemical mimicry by a bolas spider. *Science* **1977**, *198*, 1173–1175. [CrossRef]
13. Eberhard, W.G. The natural history and behavior of the bolas spider Mastophora dizzydeani sp. n.(Araneidae). *Psyche* **1980**, *87*, 143–169. [CrossRef]
14. Stowe, M.K. Prey specialization in the Araneidae. In *Spiders-Webs, Behavior, and Evolution*; Stanford University Press: Redwood City, CA, USA, 1986; pp. 101–131.
15. Yeargan, K.V. Ecology of a bolas spider, *Mastophora hutchinsoni*: Phenology, hunting tactics, and evidence for aggressive chemical mimicry. *Oecologia* **1988**, *74*, 524–530. [CrossRef] [PubMed]
16. Yeargan, K.V. Biology of bolas spiders. *Annu. Rev. Entomol.* **1994**, *39*, 81–99. [CrossRef]
17. Tanikawa, A. Japanese Spiders of the Genus *Ordgarius* (Araneae: Araneidae). *Acta Arachnol.* **1997**, *46*, 101–110. [CrossRef]
18. Diaz, C.; Long, J.H. Behavior and Bioadhesives: How Bolas Spiders, *Mastophora hutchinsoni* Catch Moths. *Insect* 2022, *submitted for review*.
19. Gertsch, W.J. The North American bolas spiders of the genera *Mastophora* and *Agatostichus*. *Bull. Am. Mus.* **1955**, *106*, 225–254.
20. Dippenaar-Schoeman, A.S.; Jocqué, R. *African Spiders: An Identification Manual*; ARC-Plant Protection Research Institute: Pretoria, South Africa, 1997; Volume 9.
21. Leroy, J.M.; Jocqué, R.; Leroy, A. On the behaviour of the African bolas-spider Cladomelea akermani Hewitt (Araneae, Araneidae, Cyrtarachninae), with description of the male. *Ann. Natal Mus. Pietermaritzbg.* **1998**, *39*, 1–9.
22. Haynes, K.F.; Yeargan, K.V.; Gemeno, C. Detection of prey by a spider that aggressively mimics pheromone blends. *J. Insect Behav.* **2001**, *14*, 535–544. [CrossRef]
23. Hutchinson, C.E. A bolas-throwing spider. *Sci. Am.* **1903**, *89*, 172. [CrossRef]
24. Shinkai, A.; Shinkai, E. The natural history, bolas construction, and hunting behavior of the bolas spider, *Ordgarius sexspinosus* (Thorell) (Araneae: Araneidae). *Acta Arachnol.* **2002**, *51*, 149–154. [CrossRef]
25. Scharff, N.; Hormiga, G. First evidence of aggressive chemical mimicry in the Malagasy orb weaving spider *Exechocentrus lancearius* Simon, 1889 (Arachnida: Araneae: Araneidae) and description of a second species in the genus. *Arthropod Syst. Phylogeny* **2012**, *70*, 107–118.
26. Hewitt, J. On certain South African Arachnida, with descriptions of three new species. *Ann. Natal Mus.* **1923**, *5*, 55–66.
27. Roff, J.; Dippenaar-Schoeman, A. Description of a new species of Cladomelea bolas-spider from South Africa, with notes on its behaviour (Araneae: Araneidae). *Afr. Invertebr.* **2004**, *45*, 1–6.
28. Chlegant, J. Kinovea (0.8.15) [Computer Software]. Joan Chlegant and Contributions. 2011. Available online: Kinveo.org (accessed on 20 December 2021).
29. Autodesk. *Autodesk Fusion 360*; Version 2.0.13881; Autodesk: San Rafael, CA, USA, 2022.
30. Autodesk. *Autodesk CFD 2023*; Version 23.0.0.0; Autodesk: San Rafael, CA, USA, 2022.
31. Fick, A. Ueber diffusion. *Ann. Phys.* **1855**, *170*, 59–86. [CrossRef]
32. Magnus, G. Ueber die Abweichung der Geschosse, und: Ueber eine auffallende Erscheinung bei rotirenden Körpern. *Ann. Phys. (Berl.)* **1853**, *164*, 1–29. [CrossRef]
33. Cardé, R.T.; Mafra-Neto, A. Mechanisms of flight of male moths to pheromone. In *Insect Pheromone Research*; Springer: Boston, MA, USA, 1997; pp. 275–290.
34. Cardé, R.T. Navigation along windborne plumes of pheromone and resource-linked odors. *Annu. Rev. Entomol.* **2021**, *66*, 317–336. [CrossRef]

35. Sahni, V.; Blackledge, T.A.; Dhinojwala, A. Viscoelastic solids explain spider web stickiness. *Nat. Commun.* **2010**, *1*, 19. [CrossRef]
36. Sahni, V.; Blackledge, T.A.; Dhinojwala, A. Changes in the adhesive properties of spider aggregate glue during the evolution of cobwebs. *Sci. Rep.* **2011**, *1*, 41. [CrossRef]
37. Elettro, H.; Neukirch, S.; Vollrath, F.; Antkowiak, A. In-drop capillary spooling of spider capture thread inspires hybrid fibers with mixed solid–liquid mechanical properties. *Proc. Natl. Acad. Sci. USA* **2016**, *113*, 6143–6147. [CrossRef]
38. Miyashita, T.; Sakamaki, Y.; Shinkai, A. Evidence against moth attraction by Cyrtarachne, a genus related to bolas spiders. *Acta Arachnol.* **2001**, *50*, 1–4. [CrossRef]

Article

Behavior and Bioadhesives: How Bolas Spiders, *Mastophora hutchinsoni*, Catch Moths

Candido Diaz, Jr. * and John H. Long, Jr.

Biology Department, Vassar College, Poughkeepsie, NY 12604, USA
* Correspondence: cdiaz@vassar.edu

Simple Summary: The bolas spider *Mastophora hutchinsoni* creates a small glue droplet attached to a web, called a bolas, which it flicks at a moth flying nearby. When it makes contact with the moth, the glue droplet soaks into the moth's scales and adheres the moth to the bolas and the spider holding it. Here, we use high-speed video to record the successful capture of moths so that we can understand the physics involved in this system and how the bolas works. In our videos, we found that the moth hovers next to the spider before being caught, minimizing the kinetic energy of the bolas upon impact. This makes the glue droplet's job much easier, giving it time to spread into the moth's scales. We noticed during capture that the glue droplet stretched like a spring as it was flicked and when it missed, it sprung back into the shape of a sphere. The glue droplets are incredibly elastic with the ability to stick to dirty surfaces. Studying their diversity, understanding their composition, and measuring their material properties are beneficial for understanding the evolution and creation of bioadhesives.

Abstract: Spiders use various combinations of silks, adhesives, and behaviors to ensnare and trap prey. A common but difficult to catch prey in most spider habitats are moths. They easily escape typical orb-webs because their bodies are covered in sacrificial scales that flake off when in contact with the web's adhesives. This defense is defeated by spiders of the sub-family of Cyrtarachninae, moth-catching specialists who combine changes in orb-web structure, predatory behavior, and chemistry of the aggregate glue placed in those webs. The most extreme changes in web structure are shown by bolas spiders, who create a solitary capture strand containing only one or two glue droplets at the end of a single thread. They prey on male moths by releasing pheromones to draw them within range of their bolas, which they flick to ensnare the moth. We used a high-speed video camera to capture the behavior of the bolas spider *Mastophora hutchinsoni*. We calculated the kinematics of spiders and moths in the wild to model the physical and mechanical properties of the bolas during prey capture, the behavior of the moth, and how these factors lead to successful prey capture. We created a numerical model to explain the mechanical behavior of the bolas silk during prey capture. Our kinematic analysis shows that the material properties of the aggregate glue bolas of *M. hutchinsoni* are distinct from that of the other previously analyzed moth-specialist, *Cyrtarachne akirai*. The spring-like behavior of the *M. hutchinsoni* bolas suggests it spins a thicker liquid.

Keywords: biomechanics; spider silk; aggregate glue; Cyrtarachninae; kinematics

Citation: Diaz, C., Jr.; Long, J.H., Jr. Behavior and Bioadhesives: How Bolas Spiders, *Mastophora hutchinsoni*, Catch Moths. *Insects* **2022**, *13*, 1166. https://doi.org/10.3390/insects13121166

Academic Editors: Thomas Hesselberg and Dumas Gálvez

Received: 29 November 2022
Accepted: 14 December 2022
Published: 16 December 2022

Publisher's Note: MDPI stays neutral with regard to jurisdictional claims in published maps and institutional affiliations.

1. Introduction

Many, but not all, species of spider use silk and varying degrees of adhesion to ensnare and trap prey long enough to envenomate them [1]. Orb-weaving spiders, known for their "wagon-wheel" shaped webs, are generalists in terms of prey, capturing a variety of aerial insects [1]. Many spider families can be distinguished by their prey capture strategies and the silk structures they create [1,2]. The most derived spiders can produce upwards of seven different silks with unique mechanical properties [1,2].

A common prey type that is difficult for most spiders to catch is the moth, since their bodies are covered in sacrificial scales that allow them to easily escape the adhesives of orb-webs [1,3,4]. Their tiny scales are weakly attached to the underlying integument, and they peel off when in contact with the adhesives of most webs [4]. This defensive mechanism works because the adhesives of orb-weaving ecribellate spiders fail to penetrate the superhydrophobic surface of scales presented by the moths [4,5]. However, these defenses have been overcome by one subfamily of spiders, Cyrtarachninae, which have evolved the ability to capture moths [3,4,6,7].

Cyrtarachninae spiders are able to catch moths because of evolutionary changes in the structure of their orb-webs and in the chemistry of their aggregate glue that is placed on those webs [3,4,8–11]. For example, species of the genus *Cyrtarachne* take the classical orb-web shape and turn it horizontal, replacing short, tight capture threads with long dangling threads [3,4,7,8]. The liquid silk, called aggregate glue, coats the capture threads and has an extremely low viscosity that allows the glue to permeate the surface of scales and spread within the matrix of channels created by the overlapping scales [4,11]. This glue penetrates not only the top layer of scales but also glues them to the cuticle below [4]. As it spreads, this glue hardens and dries, a behavior not seen in the glues of traditional orb-weaving species [4,8,9,11,12]. For this genus and a few others, the ability to catch moths comes with a trade-off in that web spinning is limited to environmental conditions with relative humidity (RH) at or above 80% [8,9,13].

The most extreme changes in the web structure of the Cyrtarachninae moth catchers are shown by the bolas spiders, who create only a single glue droplet at the end of a thread, the bolas [3,14–17]. This bolas is extremely large, several millimeters in diameter, and contains excess thread coiled within it known as a 'windlass' [14,15]. Female bolas spiders prey on male moths by releasing pheromones to draw them close; remarkably, the spiders are able alter the species of moth they are hunting throughout an evening [14,16,17]. When a moth approaches, the spider flicks its bolas, which it dangles from one of its legs, at the prey. While the unique behavior of this prey capture system has been observed in the field, the exact kinematics of the prey capture technique of bolas spiders has not been analyzed biomechanically [15–17].

Here, we use a high-speed video camera in the field to observe the kinematics of the capture behavior of *Mastophora hutchinsoni* [18]. By doing this, we hope to understand the physical and mechanical properties of the bolas during prey capture, the behavior of the moth, and how both factors lead to successful prey capture. We also use these videos to create a numerical model to explain the unique viscoelastic behavior of the bolas silk during the capture event. We observed several populations of bolas spiders throughout evenings over a week and attempted to determine if these spiders also have a humidity dependence or limitation for bolas creation.

2. Methods

2.1. Field Measurements

From 11 to 17 September 2021, the behavior of bolas spiders, *Mastophora hutchinsoni*, was observed and measured at three sites every night on the Maine Farm, University of Kentucky, Lexington, Kentucky. Each location consisted of an isolated tree either within the farmland (38.121163° N, −84.487288° W) or near the fence line directly outside of the farm (38.118291° N, −84.484114° W), (38.123160° N, −84.485876° W). Observations were made between 7 p.m. and 10:30 p.m., when spider activity ended. The order in which sites were visited varied each day. At each tree, which we identified as hackberry, *Celtis* sp., direct visual observations of bolas spiders' building behavior were recorded for a period from 10–15 min; the number of bolas spiders actively hunting (questing for prey with their front legs or creating a bolas) and the number of bolases created were tallied. During bolas creation, recordings were made of relative humidity and temperature using a hydro-thermometer (Extech model SDL500-NIST SD Logger, Extech, Nashua, NH, USA). In several instances of bolas creation, temperature and humidity were inadvertently not measured and in those cases, hourly local humidity and temperature readings were used (readings from our instrument were found to fall within ±4 %RH of these) [19]. In addition to the time spent censusing behavior, time was spent videotaping active spiders in an attempt to video the prey capture event. We also provide a Supplementary Video (SV1) we recorded from the same location in September 2022 for a different study. It shows *M. hutchinsoni's* rare trapline prey capture technique.

2.2. Kinematics of Prey Capture

Bolas spiders were observed beginning at sundown and as they transitioned from resting, to actively questing (waving their front legs in the air), to creating a bolas. Once the spider had begun to make a bolas, they were videotaped at night with a single Baslar acA1300-60 gmNIR ACE camera (Baslar AG, Ahrensburg, Germany) which was set up perpendicular to the horizontal plane of the spider. Distances between the camera and the spider varied depending on the position of the spider relative to surrounding vegetation. Prey capture events were filmed at 116 fps, the highest speed for the resolution of our camera, using a Fujinon 12.5 mm 2/3" lens (Fujifilm, Tokyo, Japan). N = 5); this sample size includes four different spiders and five capture attempts. Since most insects and arachnids do not rely on red light for vision, subjects were illuminated using an ABI LED 54 W near-infrared light (880 nm) to provide adequate lighting without impacting the behavior of the spider or moths [1].

It is important to note that we were capturing three-dimensional movements with a single camera; thus, movements out of the two-dimensional plane perpendicular from the camera resulted in measurements that underestimated the magnitudes of displacements and the velocities and accelerations they were used to derive. At the same time, at night in the field we were able to reliably capture images that resolved the droplet (Figure 1A) and the interactions of the spider and the moth (Figure 1B). From the videos, the movements of the spider, moth, and glue droplet were tracked using a combination of manual and automatic digitizing processes provided by the open-source kinematics program Kinovea (0.8.15) [20] (Figure 1). The position of the spider and moth were tracked beginning just before the moth was caught and until the spider was able to touch the moth with its front legs (N = 5). The software accurately tracked the spider due to its high contrast, but locations were manually verified. The moth was manually tracked, using the head as the focus point, because its fluttering limited the auto-tracking software.

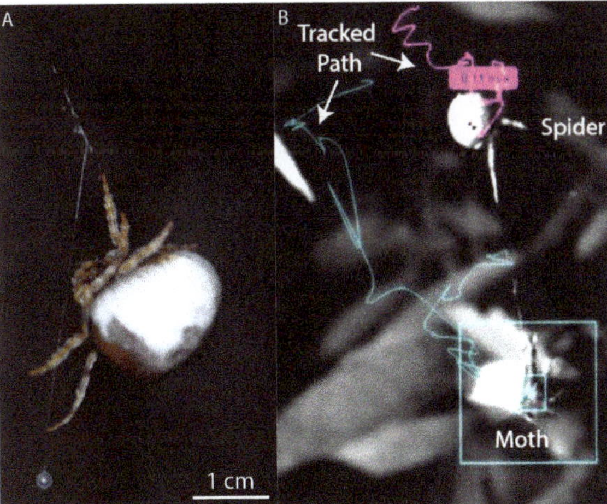

Figure 1. Nighttime, near-infrared image capture of spiders, their bolases, and the moths with which they interact. All images collected in the field. (**A**) *Mastophora hutchinsoni* dangling its bolas, with the

large glue droplet, nearly 2 mm in diameter, clearly resolved in this still image. (**B**) The moth approaches the bolas spider slowly in a path (blue) that zigzags, presenting the spider (path in purple) with a target that is close and stable. This example is typical for five of the six of the capture events recorded. It is important to note that the displacements, velocities, and accelerations that we calculated from a single high-speed camera would have underestimated the magnitudes of those properties when the motions moved out of the visual plane that was perpendicular to the camera.

From video images, the diameter of the glue droplet was measured when the droplet was still (Figure 1), and the lengths of the stretched glue droplet and radial capture stretch were measured during flicking (N = 4). A limitation of our videos is that we were only able to see the glue droplet stretch in scattered frames and only from a two-dimensional perspective, as mentioned above. We measured displacements within a two-dimensional plane and used averages over the course of the swing to estimate the forces and velocities. As a first approximation, we assumed that the motion is linear; however, there is an angular momentum to the swing of the bolas, which is likely important to the physics of the overall system but remains unaccounted for here.

The digitized displacements of the spider, bolas, and moth were used to calculate velocities and accelerations of each. Velocities were calculated using finite differences in position and the known time interval between frames. Accelerations were calculated using finite differences in the calculated velocities over the same time intervals. For all values, averages and standard deviations were calculated. Prey falling speed was calculated by measuring the slope of the prey's position over time, between being hit with the bolas and its freefall being stopped. Using these estimates of velocities and accelerations, impact and kinetic energies of the prey were calculated using the average fresh weight of the moth, 65 mg [16] and our estimate of the mass of the glue droplet from its spherical dimensions and of its density, 1.1 g cm^{-3}.

2.3. Material Properties of the Droplet

Using the measured stretching of the glue droplet during capture, and the estimated acceleration and mass of the droplet (see previous section), we estimated the glue droplet's spring constant, k (in N m^{-1}), and its damping ratio, ζ, the ratio of c, the damping coefficient, to critical damping, c_c, using a simple first-approximation physics model. We calculated k as the ratio of the maximal inertial force and the change in length of the droplet under maximal acceleration. The maximal inertial force of the droplet was estimated from $F = ma_{max}$, where m is the mass of the droplet (in kg) and a_{max} is the maximal acceleration of the droplet (ms^{-2}) along the path of the swinging thread, as measured from the video. We measured the maximal change in the diameter of the drop, Δd, as it distorted under acceleration; we recognized that as soon as the drop distorted this metric was not strictly a diameter but, rather, the longest linear dimension of the drop.

The Δd, in turn, was used to calculate a spring constant, k, for the droplet:

$$k = \frac{ma_{max}}{\Delta d} \tag{1}$$

where m is the mass of the droplet. We estimated the mass of the glue droplet as the product of the average resting diameter, d, 2 mm (average across all videos) and an estimate of the density, 1.1 g/cm^3, taken as an intermediate value between the density of water, 1.0, and the density of spider silk threads, 1.3 g/cm^3 [21].

With an estimate of k, we estimated a droplet's damping ratio, using a dynamic simulation of a linear, one-dimensional mass-spring-damper:

$$F = kd + c\dot{d} + m\ddot{d} \tag{2}$$

where d is the sinusoidally-varying diameter of the droplet with the droplet's velocity, d-dot, acceleration, and d-double-dot, as d's first- and second-order derivatives in a second-order ordinary differential equation. The damping coefficient, c, is the unknown parameter (Equation (2)). To estimate the dynamic behavior of the droplet as it is modulated by c, we modeled changes in c in the dynamic simulation of this mass-spring-damper in the Simulink environment in MATLAB (R2021b) according to established guidelines [22,23]. We varied c (μNm^{-1}) over three orders of magnitude, 0.1, 1.33, and 10, in order to examine its effect on the deformation of the droplet in three distinct dynamic states: (1) underdamped $\zeta << 1$, (2) critically damped $\zeta = 1$, and (3) overdamped $\zeta >> 1$ (for customized MATLAB code, see Supplementary Dataset S1). The denominator in ζ, c_c, is the product of 2 and the square root of the ratio of k to m; c_c was held constant by holding k and m constant at their average values (Table 1). In each of the three dynamic simulations, we measured the maximum length of the droplet as it stretched; those simulated lengths were compared to measured lengths of the droplet in order to estimate both the dynamic state of actual droplets and the droplets' value of c within an order of magnitude. Strain of the thread and droplet were measured as engineering strain.

Table 1. Kinematics of spider, moth, and bolas during prey capture.

	Average $\pm$ Standard Deviation (N = 5)
Maximum moth speed (ms^{-1})	3.75 ± 3.09
Maximum spider speed (ms^{-1})	1.44 ± 0.98
Moth Kinematics	
Impact velocity (ms^{-1})	0.22 ± 0.17
Impact kinetic energy (μJ)	2.23 ± 2.65
Maximum kinetic energy (μJ)	710.15 ± 1165.4
Capture Kinematics	
Reeling rate (ms^{-1})	0.017 ± 0.007
Falling speed (ms^{-1})	0.266 ± 0.111
Duration of reeling (s)	2.25 ± 0.34
Distance from spider when dropping (cm)	1.21 ± 0.64
Silk Kinematics	
Droplet strain (ε)	5.95 ± 1.59
Radial silk strain (ε)	0.32 ± 0.15
Droplet spring constant (μNm^{-1})	10.61 ± 4.6 (N = 4)

3. Results

In seven days, from 11 to 17 September 2021, we observed a total of ten spiders at the field site building their bolases. The following behaviors and microhabitat environmental measures were annotated from direct observations. Of those ten, four individuals were recorded capturing moths with high-speed video for a total of five events. Observations, kinematics, and modeling are reported. A single event captured as part of a different study at the same location in September 2022 is reported because it was a different type of bolas-mediated capture.

3.1. Observations on Bolas Building Behavior

During sunset, spiders began to move from their hiding spots, which were either on the underside of a leaf or on the top of a leaf with the spider camouflaged with silk splatter. As they emerged from their resting position, spiders would move along branches towards a tip, where they would build their bridge thread between two or more leaf tips or branches. Within the canopy of the tree, spiders chose positions at the crown (outer, near sun leaves) or internally (inner, near shade leaves). Because we had located spiders in their hiding spots during the day, we could determine that most spiders emerged each night; however, some remained in their resting position, sometimes with a single leg held stationary in the air. Spiders were observed from sunset until 10:30 p.m. Some spiders were observed after these times but, with little to no moth activity, those remaining spiders did not create any new bolases.

Active spiders engaged in four types of behaviors: (1) questing without a bolas, (2) creating a bolas and questing with it, (3) capturing a moth with a bolas, and (4) eating a moth. Spiders were most often found questing without a bolas, actively hanging from a thread with their front legs extended. There was no correlation between the relative humidity level near the tree and the number of active spiders (Figure 2A). However, as relative humidity increased, so too did the number of bolases that were created (Figure 2B). Please note that because individuals were sampled repeatedly, the data points are not independent statistically. Any active individual might not make a bolas, might make only one bolas, or might make multiple bolases in one evening.

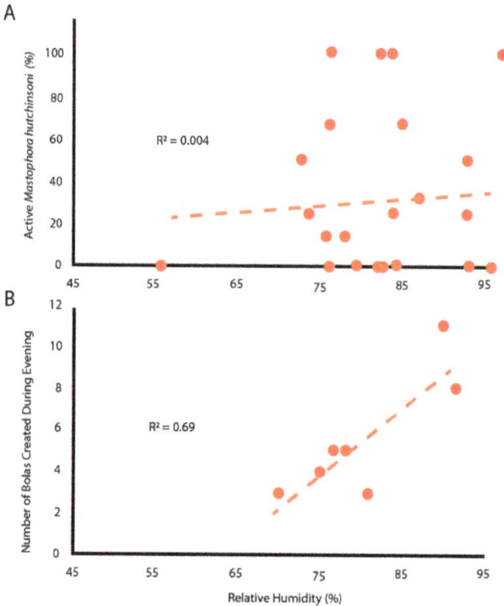

Figure 2. Humidity influenced the making of bolases but not the proportion of spiders who were active. (**A**) Relative humidity did not predict the proportion of spiders that were active. Each point represents observations of multiple spiders on a single tree during a single observation session. (**B**) Relative humidity correlates positively with the number of bolases created. Each point represents all observations on a given day.

In all of the successful moth-capturing events that we observed, the spider used a bolas to capture a bristly cutworm moth, *Lacinipolia renigera*. This does not mean that questing for a moth without a bolas is not a successful strategy, but, rather, that in our limited set of observations we did not see another method. Thus, for five of the six capture events that we observed (the sixth, a different type of bolas-mediated capture, is described in the next paragraph), a successful bolas-flicking moth-capture event can be broken down into five phases (Figure 3), which include (1) creating a bolas, (2) waiting for the moth and flicking the bolas when the moth is close, (3) resisting the escape attempted by the moth, (4) reeling in the bolas with the moth attached, and (5) subduing the moth. Drawn by pheromones produced by the spider, the male moth hovers nearby. The presence of the moth causes the spider to rapidly construct a bolas and then it waits for the next moth to approach before flicking the bolas at it. After being hit by the glue droplet at the end of the bolas, the moth executes an escape maneuver by dropping in free fall. The spider resists the attempted escape by holding onto the radial thread to which the bolas is attached while also holding onto its overhead line; it quickly reels in the attached moth, grasps it, and then subdues it by injecting venom. Reeling in the bolas is the most variable phase, as the moth may attempt to fly while the spider is reeling it in. By measuring the distance between the spider and the moth over time (Figure 3), we were able to calculate a number of kinematic parameters (Table 1).

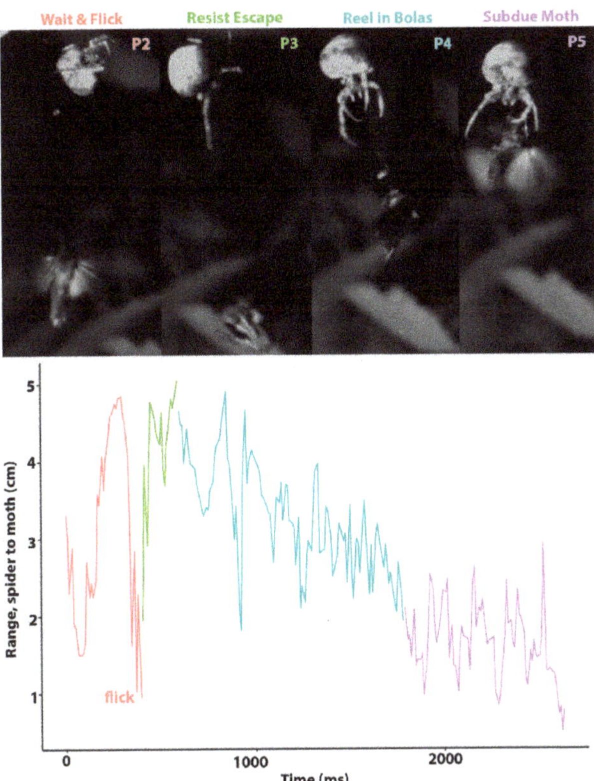

Figure 3. Bolas spider capturing a moth by flicking its bolas. In four of five successful capturing events, the spider and moth interacted in five phases, four of which are shown here (**top**). The distance between the spider and the moth, denoted as range (**bottom**), quantifies the dynamics of the struggle, with the colors of the lines corresponding to the phases above.

A dramatically different type of bolas-mediated capture event was observed and recorded once on video (Supplementary Video S1) as part of a different study on spiders at the same location but in 2022; however, it was not further analyzed. After making a single bolas in the usual manner, described above, one spider recycled the unused bolas by eating it and then moved to a new location. There, it created a nearly horizontal trapline approximately 30 cm long, from which it hung three bolases, equally spaced; the spider then moved to the end of the trapline at the higher leaf and waited. This trapline arrangement has been described by [17], but no one, to our knowledge, has observed how the architecture works during moth capture and how the spider behaves when it snares prey. On our video, a moth flew toward the spider and, on a slightly upward trajectory, was ensnared by the middle bolas. The tethered moth began to flap vigorously, spinning around the horizontal thread, causing the horizontal thread to vibrate violently as the spider moved along it toward the origin of the middle bolas. Once it reached the bolas, the spider reeled in the moth. When the moth was in the spider's grasp, it continued to flap vigorously, spinning the spider around the horizontal thread until it was fully subdued and was still. This trapline behavior differs from that described as the usual method above by substituting the first two stages—(1) creating a bolas and (2) flicking the bolas at the next moth that approaches—with (1) building a trap line, (2) moving to the end of the horizontal thread, and (3) waiting until the moth is ensnared. We also note that the acrobatic walking of the spider along the gyrating horizontal thread is a new behavioral element that is parallel in time with the moth attempting to escape.

Each bolas is composed of a radial thread and a glue droplet. The finished bolases, hanging free, have an anchoring thread that remains wrinkled/coiled, not straight, under the weight of the glue droplet. In preparation for flicking the bolas, the spider hangs by one leg from the anchoring thread and places a leg oriented below its body on the bolas near the droplet. The spider will then cock its arm, allowing the bolas to dangle down (Figure 3, 'Flick'). This is the characteristic posture that we saw in all spiders after they created a bolas and as they awaited the approach of a moth. To flick the bolas towards a nearby moth, the spider rapidly flexed the leg holding the bolas. When the flick was unsuccessful, the droplet would stretch and then recoil. When the flick was successful, attaching the droplet to the moth, the droplet stretched and remained elongated until the moth was subdued (Figure 4A–C). The maximal extent to which glue droplets stretched was, on average, 5.9 times their initial diameter (Figure 4A–C). In addition, the silk thread was also elongated, straining maximally on average 31%, which is within the range of major ampullate thread shown to stretch up to 60% [24].

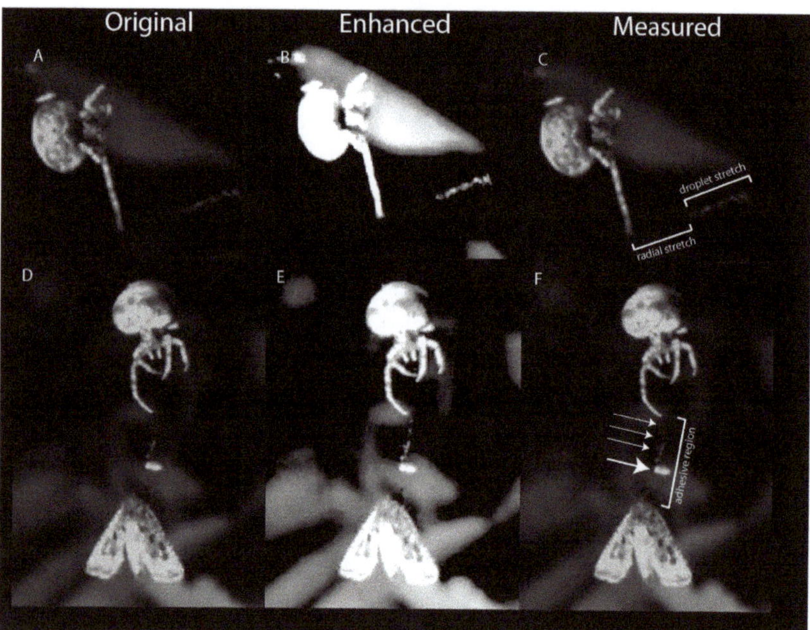

Figure 4. Behavior of the bolas during capture. The spider flicks its bolas towards the moth (off screen to the right). In the original image (**A**), a still from the high-speed video, the bolas can be seen as it stretches. Enhancement of the image (**B**), by increasing the exposure and contrast, more clearly shows the structure of the droplet as it stretches; the liquid phase of the droplet (white) remains associated with the thread of the windlass as it unfurls. The droplet continues to be tethered to the spider by the radial

thread, which also stretches (**C**). The spider reels in the moth (**D,E**). In the original image (**D**) the spider is hauling in the radial thread as the moth dangles from the stretched droplet. In the enhanced image (**E**), the scales attached to the droplet can be seen as a large clump (elliptical white structure) and smaller clumps (smaller white regions) on the windlass. The stretched droplet forms an elongated adhesive region (**F**), with a large clump (large arrow) and small clumps (small arrows) that were created during the initial deposition of the glue droplet onto the moth.

3.2. Kinematics of Prey Capture

Five of six prey capture events, when the spider flicked the bolas (see previous section), were filmed with the bristly cutworm moth, *Lacinipolia renigera*, captured in every case. In discussing the kinematics, it is important to note that we were capturing three-dimensional movements with a single camera; thus, movements out of the two-dimensional plane perpendicular from the camera resulted in measurements that underestimated the magnitudes of displacements and the velocities and accelerations they were used to derive.

The impact velocities and energies of the moths were low because the moths were hovering prior to being struck with a bolas (Table 1). Moths flew extremely close to the spider, within 2 cm, before the spider flicked the bolas. Maximum prey kinetic energy was found during thrashing and not during free fall or impact. The spiders reeled in the moths within three seconds. The maximum impact and falling energies of the moth were low (Table 1) and are observed to be well within the range of energy absorbed by spider capture threads and aggregate glue [2,25,26].

The velocities of the spider and moth varied greatly over time, with the moth having 2.6 times the maximum speed of the spider. The accelerations were highly variable as the moth thrashed and fell, with the highest magnitudes being during the moth's escape attempt while tethered. The highest velocities were also not during free fall but during escape-related thrashing (Figure 3). The velocity and acceleration of the spider are significantly lower than that of the moth, even though the two are tethered together. This can be attributed to the energy absorption of the bolas thread and the momentum fluctuations dampened by the silk line and leaf the spider is attached to.

3.3. Model of Bolas as Viscoelastic Spring during Prey Capture

The droplet of the bolas contains within it a coil of silk (the "windlass") at the center. Videos show that when the spider flicks the bolas at the moth, the mass of the glue droplet creates, by its translational and angular acceleration, a tension force on the thread, which begins to elongate (Figure 4A–C). After the thread has elongated maximally, the droplet begins to deform and reaches its maximum deformation. As the droplet deforms, the windlass inside unravels (Figure 4A–C). In videos where the spider misses its target, the deformation of the droplet is transitionary, and, after peak acceleration, it shortens in a manner that appears to be elastic. This elastic recoil is likely caused by the filamentous windlass. In contrast, the glue droplet does not recoil when the droplet impacts the moth, and the windlass continues to unwind, permitting the droplet to elongate, as the spider reels in the moth (Figure 4D–F). At the interface of the droplet and the moth, the tensile forces dislodge the moth scales and, perhaps because of their superhydrophobicity, they float to the top of the glue droplet and away from the base cuticle (Figures 4D–F and 5). Tethered to the thread, the moth, in thrashing and attempting to fly away, moves itself in a circular path. The circular path of the moth tilts the thread relative to the integument, creating an angular moment at the attachment site that may help to press the glue into the matrix of the scale. This spinning was seen in all capture events recorded on video, and it continued until the spider subdued the moth.

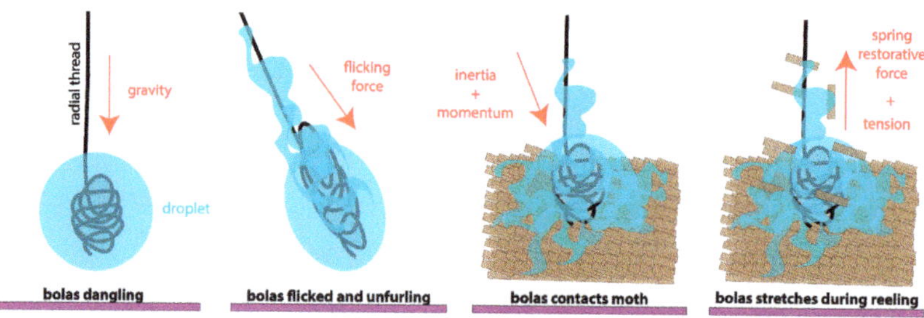

Figure 5. Speculative model of a bolas interacting with the scaled surface of the moth. The bolas has a viscous glue droplet containing a windlass of coiled silk. The flicking force of the spider stretches the glue droplet, with the internal silk of the windlass allowing it to stretch, spring-like, and to retain the liquid phase. If the droplet fails to hit the moth, the glue droplet returns elastically to its original spherical form. If the droplet hits the moth, the collision with the substrate initially dislodges scales. The tension on the thread, caused by the escaping moth, begins to unravel the windlass as it is pulled through the glue, causing the glue to spread within the matrix of the scales. The hydrophobic nature of the scales causes them to be pulled to the surface and pulled upward, cleaning the area and allowing the remaining glue to connect with the underlying cuticle. As the tethered moth struggles to escape, angular momentum is generated, leading to further contact between the glue droplet and moth substrate. The interactions of the bolas, both liquid phase and windlass, with the scales of the moth, are speculative, based on previous work of glue spreading in the previously analyzed moth-specialist, *C. akirai*.

Based on the observed dynamics of the glue droplet, we modeled its stretching during the flick as a mass-spring-damper system (Figure 6). Maximum acceleration, a_{max}, and spring constant, k, estimated from displacements measured by high-speed video and Equation (1), varied in different video observations by a factor of 4 and 3, respectively (Table 2). It is important to note that a_{max} and the k that they are used to estimate are likely underestimates, given the limitations of our two-dimensional view of a three-dimensional set of motions (see Methods). Thus, these values of a_{max} and k should be treated as rough, preliminary, order-of-magnitude estimates.

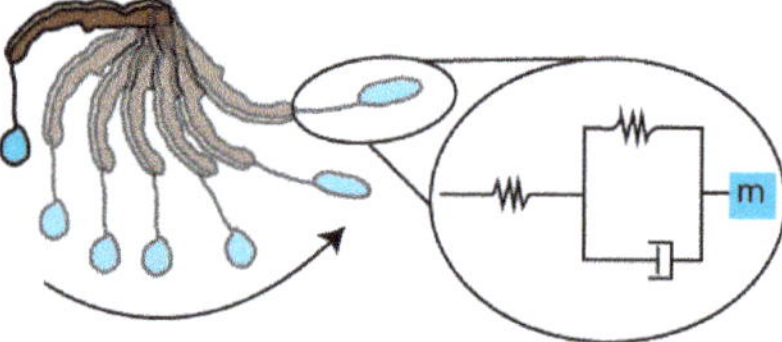

Figure 6. The dynamic behavior of the bolas during flicking, modeled as a simple mass-spring-damper system. The glue droplet is held steady with the leg in a horizontal orientation (top-most position). As the spider swings its leg, the inertia of the flicked glue droplet carries it forward until it reaches the end of its arc. At this moment, the kinetic energy is transduced into elastic energy in the radial thread and the droplet, stretching both (see Figure 4A–C). The system has two springs: (1) the radial thread and (2) the glue droplet, forming the mass-spring-damper system as in the diagram in the inset. The springs are represented by zig-zagged lines, the damper is a piston, and the center of mass is shown by the blue square, m. This model is a first-approximation and, thus, is likely to be highly simplified compared to the actual behavior of the system, which has yet to be determined.

Table 2. Bolas droplet acceleration and spring constants during a flick.

Video	Maximum Acceleration a_{max} (ms^{-2})	Spring Constant k (μNm^{-1})
1	0.103	0.6
2	0.387	1.67
3	0.449	1.72
4	0.257	1.32

Using the average k of 1.33 μNm^{-1} and average droplet mass of 4.6 mg, we modeled dynamic behavior under three different damping coefficients that were chosen to exhibit vibrations that were underdamped (damping ratio, $\zeta \ll 1$), critically damped ($\zeta = 1$), and overdamped ($\zeta \gg 1$) (Figure 7). The resulting simulations were used to qualitatively judge the droplet's actual behavior, determining which model best matched the observed motion. We aimed to determine which vibratory behavior most closely matched the droplet kinematics captured on video. With droplets stretching ~1 cm and rebounding immediately after being thrown, the closest model was the critically damped spring (Figure 7).

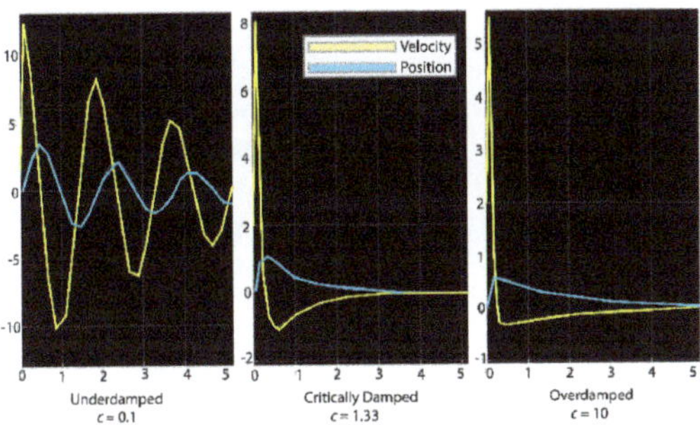

Figure 7. Mass-spring-damper estimates of bolas stretch and velocity. Underdamped $\zeta \ll 1$; critically damped $\zeta = 1$; overdamped $\zeta \gg 1$. Three models for the damping coefficient resulting in the three types of behavior. Two lines share the common y-axis with varying units; yellow lines depict estimated droplet velocity (cm s^{-1}) and blue lines the droplet's stretch from its initial position (cm). The x-axis shows time in seconds. The critically damped model yields the behavior that most closely matches actual behavior of the glue droplet. The underdamped one oscillates too much, and the glue droplet stretches too far. The overdamped does not stretch nearly as far as measured.

4. Discussion

To our knowledge, this is the first study to quantify the moth-capturing behavior of the bolas spider, *Mastophora hutchinsoni*. Using a high-speed video camera in the field, we captured the kinematics of the spider's leg and its bolas—a thread and a glue droplet—as it flicked the bolas and captured a moth. When attached to the struggling moth, the glue droplet undergoes remarkable reconfigurations, stretching to lengths nearly six times that of the droplet's original diameter. This stretching indicates that the droplet behaves as a viscoelastic spring, as indicated by our first-approximation model of the system as a critically damped mass-spring-damper. This complex mechanical behavior of the droplet is reflected in its complex composition, with a viscous liquid surrounding a coiled thread—which unspools during capture—that is continuous with the dangling thread held by the spider.

Measuring displacements over time to estimate acceleration (Table 2), and knowing the mass of the droplets, we estimated that forces involved in prey capture are relatively low, with kinetic energies of the order of 1.3 μJ. Forces are low because the moths are hovering near the droplet when they are caught; thus, the relative speed of impact of the moth and the droplet are low (Figure 1). This low impact force from hovering moths stands in contrast to the high impact speeds of fast-flying aerial prey caught in orb webs [27]. With the bolas spiders, moths generate the greatest forces after the moth drops to attempt escape and begins to thrash. When their gravity-assisted drop is arrested by the attached bolas and they are tethered, most moths respond by flying, stretching the glue droplet (Figure 4), while the spider works to quickly reel in the moth (Figure 3).

4.1. Inquisitive Prey

Male moths approach the bolas because they are attracted to pheromones produced by the spider that mimic those produced by conspecific female moths [3,16,17]. Male moths must first detect the pheromone plume and, then, to find the female, navigate up the plume's concentration gradient, a behavior that involves a zig-zag flight pattern upwind [28]. This type of chemotaxis, while common in insects, is fickle, requiring wind of low velocity, with little turbulence constant pheromone emission in order for moths to quickly locate the target [29].

However, chemotaxis for the moth is more complicated in most natural conditions, where even low-velocity winds may vary and pheromones may be released in pulses, creating odor gaps in the plume that force the moth to cast, that is, to turn perpendicular to the wind and attempt to recontact the plume [29]. The resulting flight path is slow and tortuous. Thus, as a moth approaches the bolas, it does so slowly, hovering and maneuvering as it searches for the pheromone target (Figure 1). This slow flight presents the spider with a steady target in close proximity. The flick of the bolas, when it comes, meets the moth with a low impact speed dictated more by the length of the bolas and the spider's leg than the relative speed of the moth and the spider (Figure 3). Thus, the critical first contact between glue and moth is of long duration, allowing the droplet time to permeate the scales and anchor the thread (Figure 5), which are processes that glue the moth's scales to its underlying integument.

4.2. Environmentally Constrained Predators

While specific environmental conditions are required to allow the moth to navigate by chemotaxis, the conditions must also allow the spider to make a glue droplet that stays on the tip of the thread and does not evaporate. Judging from both the activity of the spiders and the number of bolases they create (Figure 2), relative humidities above 75% appear to provide the appropriate hygroscopic balance between evaporation and absorption to allow a large droplet to be created, held in the ready for up to 30 min, and then stay attached to the web as it is flicked towards the moth. In addition, the droplets must also quickly permeate the scales of the moth, glue the scales to the moth's underlying integument, and then withstand the repeated attempts by the moth to pull free.

The apparent humidity-dependent behavior in *M. hutchinsoni* (Figure 2) may be for different reasons than the environmental constraints that affect droplet formation and mechanical properties in other species of Cyrtarachninae [11,13]. While most orb-weaver spiders make their webs and leave them for the evening, *M. hutchinsoni* glue droplets are relatively short-lived (~30 min) and are recycled by ingestion [15]. In addition, we observed that spiders did not create bolases continuously during a humid evening, nor did they make more bolases after they had successfully captured a moth. More importantly, spiders only made bolases in response to the presence of moths. Thus, moths, which are more active in high humidity [30], may directly trigger the bolas-building behavior of *M. hutchinsoni*. This, then, is an alternative hypothesis to the idea that moth-catching spiders are dependent on high relative humidity for the proper formation and function of their glue droplets [11,13]. These hypotheses may not be mutually exclusive.

In addition to considerations of relative humidity or the presence of moths, the physical structure of the local microhabitat may be important. The activity of spiders varied in ways that may indicate that their location on vegetation has an impact on behavior. Some spiders were inactive or, if active, never made a bolas. The individuals that were found farther away from the branch tips, more towards the trunk of the hackberry trees, responded first to the presence of moths, creating the largest number of bolases. Individuals located on the tips of branches showed less activity. These observations are consistent with the hypothesis that bolas spiders are responsive to variations in the amount of wind in a given tree [16]; while we did not measure wind speed, we conjecture that, in light winds, outer positions may offer the highest probability of attracting and capturing moths, while on evenings of higher winds, the inner positions may shelter or funnel wind in such a way as to allow pheromone plume formation to be coherent enough for chemotaxis by the male moths.

4.3. Predator–Prey Interactions via a Viscoelastic Bolas

Once the moth is attached to the bolas, it struggles to escape, putting dynamic loads on the bolas, stretching the droplet to lengths up to five times its original length without breaking (Table 1, Figures 2 and 4). This mechanical behavior of the droplet is remarkable for several reasons. First, the flicked droplet elongates and, if the target is missed, it will recoil; mechanical behavior that is at least partially elastic. Second, during elongation and recoil, the liquid portion of the droplet remains associated with the capture thread, which is a property of material coherence most likely related to the droplet's internal windlass, a wrapping of thread, continuous with the web from which the droplet dangles, providing attachment surface for the surrounding fluid of the droplet (Figure 5). Third, the dynamics of the droplet's deformation is consistent with the behavior of a mass-spring-damper system (Figure 6) that is critically damped (Figure 7), suggesting a matching of elastic and dissipative properties in the droplet. Finally, by undergoing extreme elongations as a critically damped elastic system, the droplet attached to the moth has, as a system, dynamic behavior that absorbs and dissipates the energy of the struggling moth. So-called "soft" springs work this way, with their low elastic modulus requiring high strain to generate a high force, slowing the rate of change in the force. What may also be important is that the droplet, by dissipating energy, prevents the moth from generating a high-magnitude jerk, where jerk is the rate of change in acceleration. The viscous dashpot of a shock absorber works this way. Both soft stiffness and high viscosity are accounted for in the mass-spring-damper viscoelastic model (Figure 6).

To help guide future studies, we offer the following prediction that should be tested for its wide phylogenetic claim. We predict that the mechanical behavior of the glue droplet during moth capture—as modeled by a simple mass-spring-damper model—developed here for M. *hutchinsoni* will also apply to the other 50 species of the genus *Mastophora*, the monophyletic taxon of bolas spiders. This prediction is based on the following assumptions: (1) all *Mastophora* species have females that capture moths by flicking a bolas, generating inertial forces in so doing that must be resisted by the droplet in order for it to stay on the web; (2) all *Mastophora* species have a windlass in their glue droplet that provides cohesive forces to retain the glue droplet as it is accelerated; and (3) all *Mastophora* species have a higher viscosity glue droplet than that found in sister taxa within the moth-catching sub-family Cyrtarachninae. These assumptions rest on incomplete data, since only a few species of *Mastophora* have been studied. A more general way to state the prediction is that any spider that flicks a bolas needs that bolas to have elastic and viscous properties that allow the droplet to (1) stay attached to the web, (2) allow the droplet to permeate the hydrophobic scales of the moth, (3) rapidly glue the scales to the underlying integument, and (4) resist the repeated dynamic forces generated by the struggling moth. In whichever form this prediction is stated, it represents the physical requirements for the behavior. Moreover, the first form of the prediction places these physical requirements into the phylogenetic context of the moth-catching specialists in the Cyrtarachninae, a taxon that has evolved a diversity of ways to catch moths, all of which involve correlated changes in web architecture, silk material properties, and behavior [12].

4.4. Adaptations for the Capture of Moths

While spiders of the Cyrtarachninae share the evolutionary innovation of catching moths, the different methods for doing so suggest that some key adaptations, shared by the common ancestor of the taxon, has permitted this rapid diversification. Making a lot of glue and having it be able to spread and harden quickly have been proposed as adaptations in the common ancestor [12]. Given the importance of being able to first permeate the super-hydrophobic scales of moths, we suspect that the glue's viscosity was an important physical property that was altered initially in the taxon's common ancestor and continued to evolve in concert with the different moth-catching behaviors seen in the extant descendent taxa. Low-viscosity glue is exemplified by that found in *Cyrtarachne akirai*, a species that catches moths with large glue droplets attached to a stationary horizontal web. Upon touching the scales of a moth, the glue droplets quickly permeate the surface of the scales and is spread by capillary forces in the micromesh created by the overlapping scales [4,11]. While in *Mastophora* viscosity has not been measured and spreading studies have not been conducted, the physical requirements mentioned above for flicking a bolas lead us to expect that they may have droplets with higher viscosity than those of *C. akirai*.

M. hutchinsoni may use the force of the bolas impacting the moth to spread the glue and force it under the scales, where capillary action can then spread it further—a trait possibly shared by other bolas spiders [31,32]. While each genus of bolas spider varies the structure and behavior of their bolas swing, all rely on the momentum generated by the spider to create impact with their prey. For example, species of the genus *Ordgarius* construct bolas of longer length than *M. hutchinsoni* that almost always contains two droplets in series. When prey is close, they spin their bolas in a circle, creating a cone-shaped space with which to strike their prey [33,34]. This genus appears less discerning to stimuli than *M. hutchinsoni* though, as they respond to human voice in addition to the wingbeats of prey, as seen in *M. hutchinsoni* [34–36]. *Cladomelea akermani* takes this indiscernibility even further and begins its prey capture technique without the stimuli of prey at all. They begin construction of a bolas of variable length, between one and four droplets, immediately at sundown. They then spin their bolas, rocking their body forward to generate and maintain momentum, for intervals of up to 15 min [37]. Thus, is seems clear that momentum of the droplet is crucial for attachment to the moth. Recent studies have shown that within superhydrophobic channels, similar to those likely to be found in the micromesh of the moth scales, fluids with higher viscosity are drawn more quickly through capillary systems than those with low viscosity [32]. The impact force pushes glue below the top of the scale line to the base cuticle creating a counter-intuitive benefit of high viscosity. This appears to be caused by air pockets between the surface roughness and the high viscosity liquid that lowers the frictional drag of the fluid as the cohesive forces of the fluid pull the bulk mass forward [32]. We look forward to future studies in Cyrtarachninae species that address the evolution of diverse biomechanical solutions for catching moths.

4.5. Caveats and Conclusions

Given our biomechanical predictions and evolutionary hypotheses, we want to acknowledge that the limited kinematic and mechanical results of this paper should be treated with caution. The kinematics are based on the 2D view of one camera; thus, any movements that are not in a plane parallel with that of the camera's optical sensor will be underestimated in terms of the magnitude, but not frequency, of displacement, velocity, and acceleration. Furthermore, the sample size is limited to ten individuals, only four of which were recorded capturing moths; one individual was recorded twice, which yielded a total of five capture events. In addition, this work was conducted in one location over an eight-day period. A small sample size and a single location may have yielded a sample of behaviors that does not represent the population-level variance of this species well. In terms of modeling the glue droplet as a viscoelastic spring, this is a first approximation based on what are likely order-of-magnitude estimates of stiffness and damping.

However, in spite of the limitations of our quantitative results, we can be certain that our recorded behavioral observations offer new insights into the hunting of *M. hutchinsoni*. The bolas spider and its target moths have, as part of their repertoire, a six-stage capture interaction that we saw repeated in five of our six recorded events (Figures 1, 3 and 4)—(1) detecting a moth, (2) creating a bolas, (3) flicking the bolas at the next moth that approaches, (4) resisting the escape attempted by moth, (5) reeling in the bolas with the moth attached, and (6) subduing the moth. The other capture event showed the spider successfully using a trapline of dangling bolases to capture a moth. Thus, this species has at least two ways of catching moths with bolases, which are (1) active flicking and (2) passive snagging. It is also worth keeping in mind that only adult females use bolases; males and early juveniles grab moths directly from the air [16]. Thus, within *M. hutchinsoni* as a whole—females, males, and juveniles—we see a variation in behaviors that is likely tied to the on-going co-evolution with the moths that they attempt to capture.

Supplementary Materials: The following supporting information can be downloaded at: https://www.mdpi.com/article/10.3390/insects13121166/s1, Video S1: Trapline Hunting technique, *Mastophora hutchinsoni,* Video S2: Moth caught on Bolas—Standard, Video S3: Moth caught on Bolas—Not Giving Up.

Author Contributions: Conceptualization, investigation, data curation, visualization, writing—original draft preparation, C.D.J.; methodology, validation, formal analysis, funding acquisition, writing—editing and revision, C.D.J. and J.H.L.J. All authors have read and agreed to the published version of the manuscript.

Funding: This work was supported by the National Science Foundation award IOS-2031962.

Data Availability Statement: Data used to create Figure 2 and Tables 1 and 2 are available as Supplementary Materials for download. The Simulink MATLAB damper-spring model and code, used to create Figure 7, are also available to download as Supplementary Materials.

Acknowledgments: We would like to thank the University of Kentucky for allowing us to use their land to locate spiders. C.D. would also like to thank Aubrey Clark for editing.

Conflicts of Interest: The authors declare no conflict of interest.

References

1. Foelix, R.F. *Biology of Spiders*; Oxford University Press: New York, NY, USA, 2011.
2. Blackledge, T.A.; Scharff, N.; Coddington, J.A.; Szüts, T.; Wenzel, J.W.; Hayashi, C.Y.; Agnarsson, I. Reconstructing web evolution and spider diversification in the molecular era. *Proc. Natl. Acad. Sci. USA* **2009**, *106*, 5229–5234. [CrossRef] [PubMed]
3. Stowe, M.K. Prey specialization in the Araneidae. In *Spiders Webs Behavior and Evolution*; Stanford University Press: Stanford, CA, USA, 1986; pp. 101–131.
4. Diaz, C.; Maksuta, D.; Amarpuri, G.; Tanikawa, A.; Miyashita, T.; Dhinojwala, A.; Blackledge, T.A. The moth specialist spider Cyrtarachne akirai uses prey scales to increase adhesion. *J. R. Soc. Interfaces* **2020**, *17*, 20190792. [CrossRef] [PubMed]
5. Andersen, S.O. Biochemistry of Insect Cuticle. *Annu. Rev. Entomol.* **1979**, *24*, 29–59. [CrossRef]
6. Miyashita, T.; Sakamaki, Y.; Shinkai, A. Evidence against moth attraction by Cyrtarachne, a genus related to bolas spiders. *Acta Arachnol.* **2001**, *50*, 1–4. [CrossRef]
7. Tanikawa, A. Two new species of the genus Cyrtarachne (Araneae: Araneidae) from Japan hitherto identified as *C. inaequalis*. *Acta Arachnol.* **2013**, *62*, 95–101. [CrossRef]
8. Cartan, C.K.; Miyashita, T. Extraordinary web and silk properties of Cyrtarachne (Araneae, Araneidae): A possible link between orb-webs and bolas. *Biol. J. Linn. Soc.* **2000**, *71*, 219–235. [CrossRef]
9. Baba, Y.G.; Kusahara, M.; Maezono, Y.; Miyashita, T. Adjustment of web-building initiation to high humidity: A constraint by humidity-dependent thread stickiness in the spider Cyrtarachne. *Naturwissenschaften* **2014**, *101*, 587–593. [CrossRef]
10. Diaz, C.; Tanikawa, A.; Miyashita, T.; Dhinojwala, A.; Blackledge, T.A. Silk structure rather than tensile mechanics explains web performance in the moth-specialized spider, Cyrtarachne. *J. Exp. Zool. Part A Ecol. Integr. Physiol.* **2018**, *329*, 120–129. [CrossRef]
11. Diaz, C.; Tanikawa, A.; Miyashita, T.; Amarpuri, G.; Jain, D.; Dhinojwala, A.; Blackledge, T.A. Supersaturation with water explains the unusual adhesion of aggregate glue in the webs of the moth-specialist spider, Cyrtarachne akirai. *R. Soc. Open Sci.* **2018**, *5*, 181296. [CrossRef]
12. Diaz, C., Jr.; Baker, R.H.; Long, J.H., Jr.; Hayashi, C.Y. Connecting materials, performance and evolution: A case study of the glue of moth-catching spiders (Cyrtarachninae). *J. Exp. Biol.* **2022**, *225* (Suppl. 1), jeb243271. [CrossRef]

13. Miyashita, T.; Kasada, M.; Tanikawa, A. Experimental evidence that high humidity is an essential cue for web building in Pasilobus spiders. *Behaviour* **2017**, *154*, 709–718. [CrossRef]

14. Eberhard, W.G. Aggressive chemical mimicry by a bolas spider. *Science* **1977**, *198*, 1173–1175. [CrossRef] [PubMed]

15. Eberhard, W.G. The natural history and behavior of the bolas spider *Mastophora dizzydeani* sp. n. (Araneidae). *Psyche* **1980**, *87*, 143–169. [CrossRef]

16. Yeargan, K.V. Ecology of a bolas spider, *Mastophora hutchinsoni*: Phenology, hunting tactics, and evidence for aggressive chemical mimicry. *Oecologia* **1988**, *74*, 524–530. [CrossRef] [PubMed]

17. Yeargan, K.V. Biology of bolas spiders. *Annu. Rev. Entomol.* **1994**, *39*, 81–99. [CrossRef]

18. Gertsch, W.J. The North American bolas spiders of the genera *Mastophora* and *Agatostichus*. *Bull. Am. Mus.* **1955**, *106*, 4.

19. Local weather forecast, news and Conditions. Weather Underground. (n.d.). Available online: https://www.wunderground.com/ (accessed on 19 September 2021).

20. Charmant, J. Kinovea (0.8.15). Joan Charmant and Contributions. 2011. Available online: https://www.kinveo.org (accessed on 27 September 2021).

21. Stauffer, S.L.; Coguill, S.L.; Lewis, R.V. Comparison of physical properties of three silks from *Nephila clavipes* and *Araneus gemmoides*. *J. Arachnol.* **1994**, *22*, 5–11.

22. Harihara, P.; Childs, D.W. *Solving Problems in Dynamics and Vibrations Using MATLAB*; Department of Mechanical Engineering Texas A & M University: College Station, TX, USA, 2014; pp. 11–21.

23. Mass-Spring-Damper in Simulink and Simscape—MATLAB & Simulink. 2021. Available online: https://www.mathworks.com/help/physmod/simscape/ug/mass-spring-damper-in-simulink-and-simscape.html (accessed on 30 September 2021).

24. Blackledge, T.A.; Cardullo, R.A.; Hayashi, C.Y. Polarized light microscopy, variability in spider silk diameters, and the mechanical characterization of spider silk. *Invertebr. Biol.* **2005**, *124*, 165–173. [CrossRef]

25. Agnarsson, I.; Blackledge, T.A. Can a spider web be too sticky? Tensile mechanics constrains the evolution of capture spiral stickiness in orb-weaving spiders. *J. Zool.* **2009**, *278*, 134–140. [CrossRef]

26. Sahni, V.; Blackledge, T.A.; Dhinojwala, A. Changes in the adhesive properties of spider aggregate glue during the evolution of cobwebs. *Sci. Rep.* **2011**, *1*, 1. [CrossRef]

27. Sensenig, A.T.; Lorentz, K.A.; Kelly, S.P.; Blackledge, T.A. Spider orb webs rely on radial threads to absorb prey kinetic energy. *J. R. Soc. Interfaces* **2012**, *9*, 1880–1891. [CrossRef] [PubMed]

28. Cardé, R.T.; Mafra-Neto, A. Mechanisms of flight of male moths to pheromone. In *Insect Pheromone Research*; Springer: Boston, MA, USA, 1997; pp. 275–290.

29. Cardé, R.T. Navigation along windborne plumes of pheromone and resource-linked odors. *Annu. Rev. Entomol.* **2021**, *66*, 317–336. [CrossRef] [PubMed]

30. Wolfin, M.S.; Raguso, R.A.; Davidowitz, G.; Goyret, J. Context-dependency of in-flight responses by Manduca sexta moths to ambient differences in relative humidity. *J. Exp. Biol.* **2018**, *221*, jeb.177774.

31. De Gennes, P.G.; Brochard-Wyart, F.; Quéré, D. *Capillarity and Wetting Phenomena: Drops, Bubbles, Pearls, Waves*; Springer: New York, NY, USA, 2004; Volume 315.

32. Vuckovac, M.; Backholm, M.; Timonen, J.V.I.; Ras, R.H.A. Viscosity-enhanced droplet motion in sealed superhydrophobic capillaries. *Sci. Adv.* **2020**, *6*, eaba5197. [CrossRef] [PubMed]

33. Tanikawa, A. Japanese Spiders of the Genus Ordgarius (Araneae: Araneidae). *Acta Arachnol.* **1997**, *46*, 101–110. [CrossRef]

34. Shinkai, A.; Shinkai, E. The natural history, bolas construction, and hunting behavior of the bolas spider, *Ordgarius sexspinosus* (Thorell) (Araneae: Araneidae). *Acta Arachnol.* **2002**, *51*, 149–154. [CrossRef]

35. Haynes, K.F.; Yeargan, K.V.; Gemeno, C. Detection of prey by a spider that aggressively mimics pheromone blends. *J. Insect Behav.* **2001**, *14*, 535–544. [CrossRef]

36. Hutchinson, C.E. A bolas-throwing spider. *Sci. Am.* **1903**, *89*, 172. [CrossRef]

37. Leroy, J.; Jocquac, R.; Leroy, A. On the behaviour of the African bolas-spider *Cladomelea akermani* Hewitt (Araneae, Araneidae, Cyrtarachninae), with description of the male. *Ann. Natal Mus.* **1998**, *39*, 1–9.

Brief Report

Novel Observation: Northern Cardinal (*Cardinalis cardinalis*) Perches on an Invasive Jorō Spider (*Trichonephila clavata*) Web and Steals Food

Arty Schronce [1] and Andrew K. Davis [2],*

[1] Independent Researcher, 263 Berean Avenue, Atlanta, GA 30316, USA
[2] Odum School of Ecology, University of Georgia, Athens, GA 30602, USA
* Correspondence: akdavis@uga.edu

Simple Summary: A spider native to east Asia (jorō spider) is spreading in the United States, and a pressing question is how it will affect native fauna. This brief report details an observation of a bird that foraged for food from a jorō web, all while perching on it. This demonstrates one small (positive) impact of the spider, and also emphasizes just how strong its webs are. Additional experimental data on web strength confirms that the webs are capable of supporting a similarly-sized songbird.

Abstract: An invasive spider (*Trichonephila clavata* [L. Koch 1878], or jorō spider) is rapidly expanding throughout the southeast of the United States, engendering many questions about how native fauna will be affected. Here, we describe an observation of a northern cardinal (*Cardinalis cardinalis*, L.) consuming prey items from a jorō web, which serves as an example of a native species deriving a (small) benefit from this new invader. Moreover, the manner of the kleptoparasitism is also noteworthy; the cardinal perched directly on the web, which supported its weight (which is 42–48 g in this species). This appears to be the first documented case of a spider web supporting a perching bird. We also include measurements of other jorō webs, where web strength had been assessed using a force gauge, which revealed that typical webs can support masses up to 70 g before collapsing. Collectively, this information adds to the small but growing body of knowledge about the biology of this non-native spider.

Keywords: *Trichonephila clavata*; jorō spider; invasive; northern cardinal; web strength

Citation: Schronce, A.; Davis, A.K. Novel Observation: Northern Cardinal (*Cardinalis cardinalis*) Perches on an Invasive Jorō Spider (*Trichonephila clavata*) Web and Steals Food. *Insects* **2022**, *13*, 1049. https://doi.org/10.3390/insects13111049

Academic Editors: Thomas Hesselberg and Dumas Gálvez

Received: 19 October 2022
Accepted: 10 November 2022
Published: 13 November 2022

1. Background

Spiders make webs to procure food for themselves, although these same webs, and their trapped prey, are sometimes exploited by other species. For example, species of *Argyrodes* spiders reside on or near webs of other orb-weavers and consume trapped insects (a behavior known as kleptoparasitism) [1,2]. Certain species of predatory fireflies have been observed stealing trapped fireflies from spider webs, to sequester their chemical defenses [3]. There are species of hover wasps that consume trapped prey in orb webs [4]. Further, avian kleptoparasitism of insects from spider webs has also been documented; Waide and Hailman [5] described multiple reports of birds that were observed hovering next to spider webs while gleaning the trapped insects (or insect carcasses). These reports included a bunting, a vireo, a warbler, a wren and a hummingbird. Parrish [6] also described instances of hummingbirds stealing food from spider webs in Utah. In addition to this published literature, there are a variety of anecdotal observations and videos on the internet of various songbirds stealing prey items from spider webs. We note that throughout all of these published and anecdotal cases, the birds in question were observed hovering near the webs or perched on branches nearby. The following report describes an observation of a bird native to North America procuring food from a non-native spider web in a very unusual way, by perching directly on the web.

Jorō spiders, *Trichnophila clavata* L. Koch 1878 (Figure 1A), are an orb-weaving species native to Japan and eastern Asia but have recently been introduced to the southeast United States, being fist observed in 2013 in a few locations in northern Georgia [7] and are now expanding their range. They are expected to continue spreading beyond the southeast, since their physiology appears suited for surviving the colder climates of the north of the United States [8]. This spider species is large (with outstretched legs, up to 10 cm), and they have a striking color pattern of black, yellow and red (Figure 1A). Importantly, their webs are very conspicuous; they are typically up to 1 m in diameter, with a three-dimensional structure (Figure 1B). In the new United States region, these webs are often found in human-dominated landscapes, such as urban areas, and can be built on a wide range of human structures or close vegetation (Davis, pers. obs.). Additionally, of note, is that the individual web fibers are exceptionally strong; spiders in this genus are known for producing silk with high tensile strength [9–11]. The following observation also speaks to this exceptional fiber strength.

Figure 1. (**A**) A female jorō spider, *Trichnephila clavata*, in its web. Note the complex (not two-dimensional) web pattern. Photo taken by A. Davis in Watkinsville, GA on 18 October 2020. (**B**) Photograph of the jorō spider web in Atlanta where the observation occurred. The web was strung between stalks of wintersweet and a neighboring structure and was approximately 1.25 m × 1.25 m in size.

2. Observation

The observation in question took place at the residential home of the first author, in Atlanta, GA, which is an area where jorō spiders have successfully expanded into. There, a jorō spider had built a web next to the side of the house facing the neighboring house, and it had used stalks of wintersweet as support on one side, and the neighboring house on the other (Figure 1B). Based on measurements by the author, this web was approximately 1.25 m × 1.25 m in size, and it was approximately 2 m off the ground. From anecdotal observations of other jorō webs, this web size, structure, and placement appears to be typical of the species in its new range (Davis, pers. obs.). Importantly, jorō spider webs are not completely circular, but tend to have a flattened or level section on the top edge, as is visible in Figure 1B. Presumably, these threads provide structural support.

On 13 September 2022, the first author observed a female cardinal (*Cardinalis cardinalis*, L.) perched on the top support strands of the web (the web was visible from a screen door). The author did not witness at what point it alighted, but presumably it was not long before the observation. At first, the author believed the bird had been trapped, but this turned out not to be the case. The bird was in fact perched on the web itself (not a branch or support structure), and near the middle of the web, which appeared capable of supporting the bird. The author made sure this was the case by watching closely, and also by examining the web afterward. The author was able to take photos of the perched bird through a screen door (Figure 2). The author observed the cardinal lunge toward the spider (while perched), though the spider moved away from the bird. It is unclear if this was an attempt to capture the spider or to warn it off.

Figure 2. Kleptoparasitism behavior by a northern cardinal (*Cardinalis cardinalis*), observed on 13 September 2022 in Atlanta, GA (Fulton Co.). The cardinal perched on the top of the jorō spider web (which did not break) and proceeded to pick off trapped insects from the web. The spider (arrow) moved to the edge of the web during the encounter. The bird flew off after ~2 min, seemingly without effort or entanglement. The web remained intact and was present the following day. Photo taken by A. Schronce (through a screen door).

Next, the cardinal proceeded to "glean" from the web, by pecking at (and eating) discarded insect carcasses and/or trapped prey items, all while still perched on the top strands. This behavior lasted for approximately 2 min. Then, the cardinal flew off the web, seemingly without effort or entanglement. After closer inspection, the web itself was not apparently damaged from this event, and the spider was observed in the (undamaged) web the next day. As of the time of this writing (14 October 2022), the spider and web were still present.

3. Additional Data in Support of the Observation

To help substantiate the observation above (of a jorō web supporting the weight of a northern cardinal), one of the authors (Davis) drew upon a previously conducted, unpublished set of measurements of actual jorō web strength. In the fall of 2021, the author and an assistant had measured some of the many jorō webs surrounding and/or near his own home in Oconee County, GA, as part of another project. A total of 10 webs of similar size as the Atlanta web had been selected (approximately 1 m × 1 m). For each measurement, a fine thread was looped over a given web (near the midpoint) so that it encircled the web, then the bottom end of the thread loop was attached to an electronic force gauge (Pasco Passport Force Sensor, Pasco.com; Figure 3). The gauge was pulled downward until the web broke; the point at which it broke was recorded, in Newtons, on a laptop computer. Note that this measurement was not an index of individual fiber tensile strength, but rather a metric of the downward force that a bird would apply on the collective web, if perched in the middle (as the cardinal did). From measurements of 10 webs, the average force required to collapse the web was 0.68 N (sd = 0.42, range: 0.2 to 1.5 N), which is equivalent to a downward weight of 69 g.

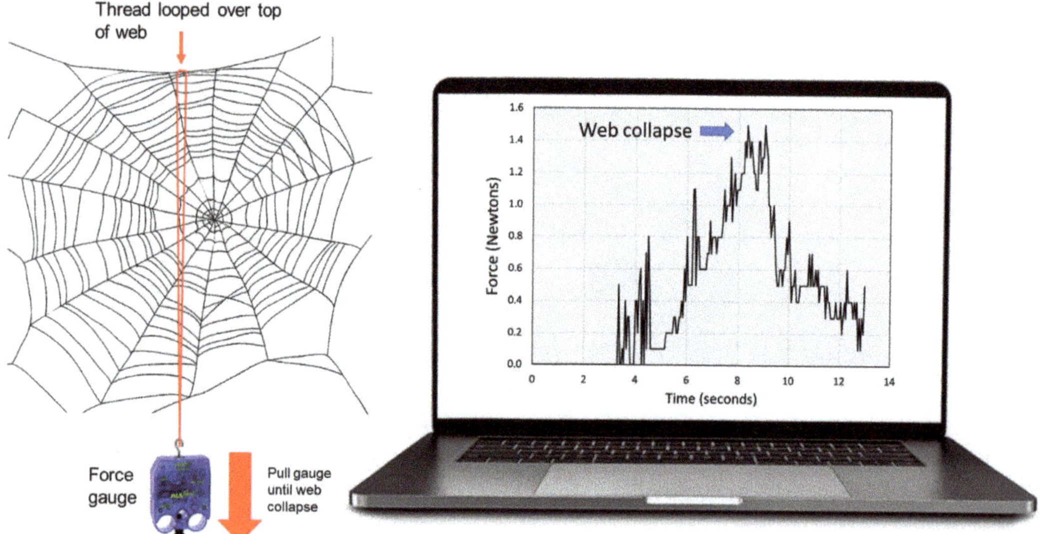

Figure 3. Schematic diagram showing how jorō web strength was measured, on 10 webs in Oconee Co., Georgia. One of the authors (Davis) looped a light thread around the middle of a given web, then tied the lower end of the loop to a handheld, digital force gauge. The gauge was pulled downward until the web collapsed, and the peak force (in Newtons) at which the web broke was recorded on a laptop. Raw force data from one web test is shown for illustration.

4. Discussion

There are two interesting elements of this observation, which should each add to the small but growing body of knowledge of the biology of the newly invasive jorō spider in the United States. First, the fact that a full-sized northern cardinal (which typically weighs 42–48 g, [12]) could perch on this spider's web without it breaking appears to be a scientific first; to our knowledge, this represents the first documented case of a spider web (of any species) supporting a perching bird, and it underscores just how strong the webs of this particular species are. In fact, the additional data provided on jorō web strength does confirm that an average web should be capable of supporting a bird of that size.

Spiders in the genus *Trichonephila* have long been studied for the biomechanical properties of their silk, which is exceptionally strong [9–11,13–16]. Throughout this body of literature, we note that the means of measuring silk strength has usually involved testing tensile strength of individual silk fibers using lab-based machinery (e.g., [15–17]), which is no doubt necessary for accuracy and repeatability. However, these individual fiber measurements are not easily extrapolated to provide estimates of whole-web strength, since webs of different species (and individuals) can vary in complexity, size, and number of support strands. It is possible to evaluate whole-web mechanical strength using mathematical models [18], which is beyond the scope of this paper, but such a task would be daunting with *T. clavata*, given the complexity of their webs. Moreover, in some species there is even variation in fiber strength and elasticity across the different parts of web, from the fibers that make up the outer frame, to the inner catchment spiral [19]. We further note that the complex, 3-dimensional structure of jorō webs, in particular (with many supporting and anchoring threads), may actually enhance their web strength even further than their fiber strength provides. Note that the web in question (Figures 1B and 2) has a complex design with a variety of support strands.

The other noteworthy aspect of this observation is that it demonstrates at least one way in which a native species could receive a small, but positive benefit from this newly invasive spider; in this one instance, the cardinal apparently utilized the jorō web as a one-time food resource. Cardinals have not previously been reported to eat from spider webs [5]; perhaps in the presence of the sheer numbers of these new webs, along with their large size and abundance of trapped prey, birds, such as cardinals, may learn to exploit this resource. Interestingly, one of the authors (Davis) has also observed native dewdrop spiders (genus *Argyrodes*) in jorō spider webs around his home; these spiders are known kleptoparasites, and even of jorō spiders in their native range [1]. The extent to which the jorō spider webs in North America become hosts to dewdrop spiders deserves further study, as does the use of their webs by native birds.

The jorō spider is rapidly expanding its range in the southeast of the United States, and will likely continue throughout a large portion of the United States, based on its physiology [8]. Therefore, understanding how it will impact the native fauna (positively or negatively) is a priority for research. Such studies are currently underway but are only in their infancy. The information presented in this report should help to further this overarching goal of understanding how this non-native spider will impact the native ecosystem of North America.

Author Contributions: A.K.D.—data collection, manuscript writing. A.S.—data collection, reviewing manuscript. All authors have read and agreed to the published version of the manuscript.

Funding: This research received no external funding.

Data Availability Statement: The force measurements presented in this study are available on request from the corresponding author.

Acknowledgments: We thank Oscar Davis for help collecting measurements of web strength. The manuscript was improved after helpful comments from three anonymous reviewers.

Conflicts of Interest: The authors declare no conflict of interest.

References

1. Miyashita, T. Competition for a limited space in kleptoparasitic *Argyrodes* spiders revealed by field experiments. *Popul. Ecol.* **2001**, *43*, 97–103. [CrossRef]
2. Agnarsson, I. Spider webs as habitat patches—The distribution of kleptoparasites (*Argyrodes, theridiidae*) among host webs (Nephila, tetragnathidae). *J. Arachnol.* **2003**, *31*, 344–349. [CrossRef]
3. Faust, L.; De Cock, R.; Lewis, S. Thieves in the night: Kleptoparasitism by fireflies in the genus *Photuris* Dejean (Coleoptera: Lampyridae). *Coleopt. Bull.* **2012**, *66*, 1–6. [CrossRef]

4. Rossler, D.C.; Ogan, S.; Curio, E.; Krehenwinkel, H. Ability makes a thief: Vision, learning, and swift escape help kleptoparasitic hover wasps not to fall prey to their spider hosts. *Behav. Ecol. Sociobiol.* **2019**, *73*, 152. [CrossRef]
5. Waide, R.B.; Hailman, J.P. Birds of five families feeding from spider webs. *Wilson Bull.* **1977**, *89*, 345–346.
6. Parrish, J.R. Kleptoparasitism of insects by a broad-tailed hummingbird. *J. Field Ornithol.* **1988**, *59*, 128–129.
7. Hoebeke, E.R.; Huffmaster, W.; Freeman, B.J. *Nephila clavata* L Koch, the Joro Spider of East Asia, newly recorded from North America (Araneae: Nephilidae). *PeerJ* **2015**, *3*, e763. [CrossRef] [PubMed]
8. Davis, A.K.; Frick, B.L. Physiological evaluation of newly invasive joro spiders (*Trichonephila clavata*) in the southeastern USA compared to their naturalized cousin, *Trichonephila clavipes*. *Physiol. Entomol.* **2022**, *2022*, 1–6. [CrossRef]
9. Higgins, L.; Rankin, M.A. Nutritional requirements for web synthesis in the tetragnathid spider *Nephila clavipes*. *Physiol. Entomol.* **1999**, *24*, 263–270. [CrossRef]
10. Kitagawa, M.; Kitayama, T. Mechanical properties of dragline and capture thread for the spider *Nephila clavata*. *J. Mater. Sci.* **1997**, *32*, 2005–2012. [CrossRef]
11. Moon, M.J. Fine structure of the aggregate silk nodules in the orb-web spider *Nephila clavata*. *Anim. Cells Syst.* **2018**, *22*, 421–428. [CrossRef] [PubMed]
12. Halkin, S.L.; Shustack, D.P.; De Vries, M.S.; Jawor, J.M.; Linville, S.U. Northern Cardinal (*Cardinalis cardinalis*), version 2.0. In *Birds of the World*; Poole, A.F., Gill, F.B., Eds.; Corenll Lab of Ornithology: Ithaca, NY, USA, 2021. [CrossRef]
13. Ebenstein, D.M.; Wahl, K.J. Anisotropic nanomechanical properties of *Nephila clavipes* dragline silk. *J. Mater. Res.* **2006**, *21*, 2035–2044. [CrossRef]
14. Rousseau, M.E.; Cruz, D.H.; West, M.M.; Hitchcock, A.P.; Pezolet, M. *Nephila clavipes* spider dragline silk microstructure studied by scanning transmission X-ray microscopy. *J. Am. Chem. Soc.* **2007**, *129*, 3897–3905. [CrossRef] [PubMed]
15. Jiang, P.; Guo, C.; Lv, T.Y.; Xiao, Y.H.; Liao, X.J.; Thou, B. Structure, composition and mechanical properties of the silk fibres of the egg case of the Joro spider, *Nephila clavata* (Araneae, Nephilidae). *J. Biosci.* **2011**, *36*, 897–910. [CrossRef] [PubMed]
16. Gosline, J.M.; Demont, M.E.; Denny, M.W. The structure and properties of spider silk. *Endeavour* **1986**, *10*, 37–43. [CrossRef]
17. Zax, D.B.; Armanios, D.E.; Horak, S.; Malowniak, C.; Yang, Z.T. Variation of mechanical properties with amino acid content in the silk of *Nephila clavipes*. *Biomacromolecules* **2004**, *5*, 732–738. [CrossRef] [PubMed]
18. Yu, H.; Yang, J.L.; Sun, Y.X. Energy absorption of spider orb webs during prey capture: A mechanical analysis. *J. Bionic Eng.* **2015**, *12*, 453–463. [CrossRef]
19. Gosline, J.M.; Guerette, P.A.; Ortlepp, C.S.; Savage, K.N. The mechanical design of spider silks: From fibroin sequence to mechanical function. *J. Exp. Biol.* **1999**, *202*, 3295–3303. [CrossRef] [PubMed]

Brief Report

Tangled in a Web: Management Type and Vegetation Shape the Occurrence of Web-Building Spiders in Protected Areas

El Ellsworth [1,†], Yihan Li [1,†], Lenin D. Chari [2], Aidan Kron [1] and Sydney Moyo [1,*]

1   Department of Biology and Program in Environmental Studies and Sciences, Rhodes College, Memphis, TN 38112, USA
2   Department of Zoology and Entomology, Rhodes University, Makhanda 6140, South Africa
*   Correspondence: moyos@rhodes.edu
†   These authors contributed equally to this work.

Simple Summary: Spiders are among the most common predators in terrestrial ecosystems and play a crucial role in ecosystems. However, with changing environments, spiders are under pressure from pollution and habitat destruction. In this study, we collected spiders from five parks with different management histories in the greater Memphis, Tennessee area to explore the extent to which human oversight and management of natural areas, especially invasive plant management, influence spider occurrence. Our results showed that invasive plants might provide a valuable habitat for the humpbacked orb-weaver, which was predominantly found on invasive plant species. These findings may have implications for the management of invasive plants in parks and other protected areas.

Abstract: Land management of parks and vegetation complexity can affect arthropod diversity and subsequently alter trophic interactions between predators and their prey. In this study, we examined spiders in five parks with varying management histories and intensities to determine whether certain spider species were associated with particular plants. We also determined whether web architecture influenced spider occurrence. Our results showed that humpbacked orb-weavers (*Eustala anastera*) were associated with an invasive plant, Chinese privet (*Ligustrum sinense*). This study revealed how invasive plants can potentially influence certain spider communities, as evidenced by this native spider species only occurring on invasive plants. Knowing more about spider populations—including species makeup and plants they populate—will give insights into how spider populations are dealing with various ecosystem changes. While we did not assess the effect of invasive plants on the behavior of spiders, it is possible that invasive species may not always be harmful to ecosystems; in the case of spiders, invasive plants may serve as a useful environment to live in. More studies are needed to ascertain whether invasive plants can have adverse effects on spider ecology in the long term.

Keywords: invasive; tennessee; architecture; management; plants; IndVal

Citation: Ellsworth, E.; Li, Y.; Chari, L.D.; Kron, A.; Moyo, S. Tangled in a Web: Management Type and Vegetation Shape the Occurrence of Web-Building Spiders in Protected Areas. *Insects* **2022**, *13*, 1129. https://doi.org/10.3390/insects13121129

Academic Editors: Thomas Hesselberg and Dumas Gálvez

Received: 28 October 2022
Accepted: 3 December 2022
Published: 7 December 2022

Publisher's Note: MDPI stays neutral with regard to jurisdictional claims in published maps and institutional affiliations.

1. Introduction

The role of biological invasions has been emphasized by many conservationists, with many of the findings showing that biological invasions can have striking ecological and socio-economic consequences [1–3]. Invasive alien plants are known to be one of the major drivers of global biodiversity declines [4,5]. In addition to potentially causing biodiversity declines, invasive plant eradication can be a costly endeavor for land managers [6].

At the epicenter of biological invasions are anthropogenic-related stressors. For instance, anthropogenic activities such as urbanization and globalization have contributed to the influx of invasive species in ecosystems globally, particularly after the Industrial Revolution and in the modern era [7]. Globally, 37% of all first records of invasive species were reported between 1970–2014 [7]; Europe alone had more invasive plant, mammal, and invertebrate introductions in 1975–2000 compared to any other time after the 16th century [8]. Such invasive species tend to displace endemic species through competition and can create noticeable changes in environments and ecosystem dynamics, particularly deleterious ones, which has greatly concerned scientists [9]. For instance, invasive species in Australia caused huge declines in native species populations [10]. Consequently, these invasions are of foremost importance and concern, as they disrupt the ecosystem balance maintained by native species. Plants in particular are of significant importance in ecosystems, as they provide many ecosystem services including food resources, hiding spots, and micro-habitats. As a result, when foreign plants are introduced, they may cause cascading effects that affect different levels of an ecosystem, including animal habitats, species distributions, and ecological roles. For example, native plants in Rwanda are important in preserving ideal physicochemical soil properties for soil-litter arthropods, supporting greater arthropod abundance compared to exotic plants [11].

Several researchers have studied the ways to mitigate the negative impacts of invasive plants through invasive species management (ISM), which costs billions of dollars annually [12]. However, it is worth noting that biological invasions are not straightforward issues. Invasive species management interacts with social, economic, and environmental factors, and if carried out improperly, it can cause unintended consequences [13]. One such effort was in a Hawaiian lowland wet forest where managers removed invasive plants over a decade to promote native biodiversity and return the forest to its pre-invaded state [14]. Subsequently, the removal of invasive plants unintentionally recruited even more invasive species [14]. Situations are further complicated in areas with well-established invasive species. For example, some research suggests that interdependency has emerged between local avian frugivore species and the invasive *Lonicera* species in Central Pennsylvania, USA, and direct removal may cause undesirable decrease in frugivore abundance [15]. Certain species have successfully adapted to environmental changes caused by invasive plants. For example, while invasive plants can have negative impacts on herbivores due to unpalatability, they provide greater structural complexity for various invertebrates [16–18]. Inherent biases might cause people, including scientists, to categorize invasive species and their impact on native ecosystems as singularly bad, despite demonstrations that natural invasions rarely caused notable ecological damage or reduced species richness [19–21]. As such, many scholars have suggested looking at introductions of invasive species as accidental experiments and quantifying their subsequent impacts in an unbiased way as to fully explore how ecosystem dynamics change when invasives establish themselves [22]. The aforementioned studies reveal the complex interactions between invasive plants and terrestrial organisms. To fully understand the extent of invasive species on environments, further research is needed on how invasives impact specific wildlife species in specific areas [23].

Invertebrates, specifically spiders, may serve as good biological indicators of overall ecosystem health for several reasons. First, with an annual prey kill of 400-800 million tons (compared to around 400 million tons of annual human meat and fish consumption), spider communities play important roles in the terrestrial ecosystem and ecosystem dynamics worldwide [24]. Second, because many spiders are habitat specialists that react to environmental changes and stress [25,26], they may be good model organisms to study habitat perturbations. Lastly, spider diversity is usually not related to plant species numbers [25,27,28] but rather to the structure (e.g., number of contact points) and microclimate of the habitat [29,30], which makes vegetation structure a good predictor of spider community occurrence. Because web-building spiders (e.g., the families Araneidae, Nephilidae, Tetragnathidae, Theridiidae, Uloboridae, Linyphiidae) often use plants as a substrate to spin their webs, their fitness and role in ecosystems can rely on plants [31,32]. This makes spiders a good model system to explore some of the impacts that invasive plants may have on terrestrial ecosystems.

To gain insights into the effects of human management of invasive plant species on local ecosystem, we examined how spiders, one of the most ubiquitous predators, interact with environments of varying plant distributions and ISM histories. To this end, we addressed the following guiding questions: (1) How do web characteristics and prey types relate to spider species occurrence? (2) Which parks and trees are predominantly associated with select species of spiders? (3) How is spider ecology potentially impacted by differing invasive plant species distributions?

2. Materials and Methods

2.1. Study Area and Site Descriptions

Specimens were collected from five parks within the greater Memphis, Tennessee area (Figure 1). We selected these five parks because of the heterogeneity in their management histories (Table 1). It was assumed that different management practices would translate into differences in plant structure and communities. The tree species at these parks (Figure 1) have been adequately characterized based on the dominant vegetation (Laport, unpublished notes), making the endeavor in our study feasible. In addition, we selected sites within parks with previously characterized vegetation communities (Laport, unpublished notes). Climate, rainfall, and seasons are described elsewhere [33,34].

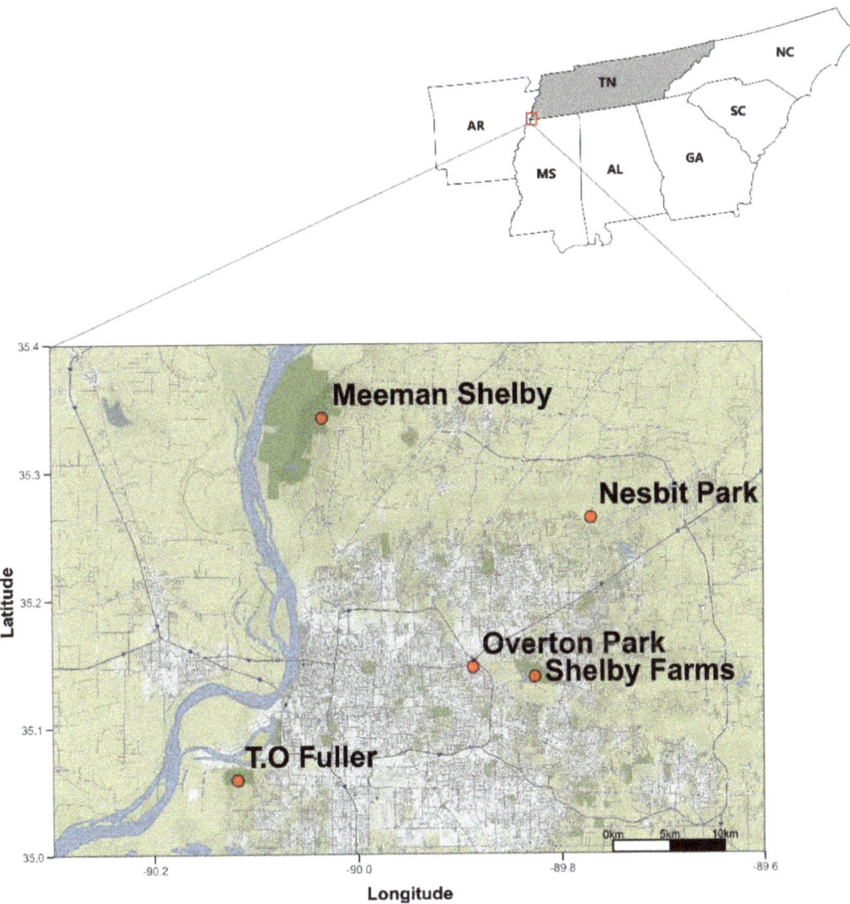

Figure 1. Map of Memphis showing five parks sampled during this study. Samples were collected in Memphis, Tennessee, USA. Maps and drawings were drawn using the 'ggmap' package in R [35].

Table 1. Site descriptions for samples collected in Memphis, TN, Summer 2022. Shelby Farms and Overton have more overall current funding, and consequently more management and human oversight, than Nesbit, Meem.an-Shelby, and T.O. Fuller [36–38].

Site	Description	Dominant Plants	Invasive
Shelby Farms Park	4500-acre public park that was opened for public recreation by the Shelby County government in the 1970s; is heavily managed, including forest trail upkeep.	Winged Elm Slippery Elm Privet	No No Yes
Overton Park Conservatory	184-acre public park opened in 2011 with heavy maintenance, including invasive species removals and trail upkeep.	Hornwort Pawpaw Box Elder Sweetgum	No No No No
Nesbit Park	333.75-acre park owned by cycling club company Stanking Creek Cycling open for free public use. No explicit plant management other than trail upkeep.	Jumpseed Pawpaw Winged Elm Sweetgum	No No No No
Meeman-Shelby State Park	13,469-acre state park initially built in the 1930s by the National Park Service; had land clearing and replanting efforts when built but currently has less spending management than Shelby Farms or Overton Parks.	Hophornbeam Beech Hophornbeam Sugar Maple	No No No No
T.O. Fuller State Park	1138-acre state park opened in 1938; has public recreation options but significantly lower revenues and spending than Shelby Farms and Overton Parks in recent years.	Pawpaw Privet Winged Elm Slippery Elm	No Yes No No

2.2. Sample Collections and Laboratory Processing

Spider collections were restricted to spiders found within webs in the field. In theory, this was an effort to predominantly collect specimens from web-building spider families (e.g., Araneidae, Tetragnathidae, Theridiidae and Linyphiidae) which utilize plants as a web anchor. Because a number of spiders rarely crawl on the ground and are restricted to their webs and the tree branches, active samplings via aerial hand collections were the most appropriate and conservative sampling method [39]. Active sampling is considerably less destructive to invertebrate microhabitats relative to methods such as beating and fogging that disturb or kill non-target invertebrates [40,41]. This was important to minimize our footprint on the environment during our study and preserve invertebrate biodiversity. Spiders were therefore collected using the hand-to-jar technique [42]. Samplings were conducted on a catch-per-unit-area basis to gain information on different habitats. Time and effort were standardized in each transect twice (beginning of the summer and end of the summer of 2022), and the same transects were used in subsequent sampling trips. In each season, at each site, intensive searches for web-building spiders were performed by two collectors within each transect. All collections took place between 06h00 and 12h00. Due to the difficulty in collecting spiders in out-of-reach heights, searches were restricted to a maximum vertical height of 2 m. Collections were conducted on climatically stable days (i.e., clear days without heat advisories or rainfall). A commercial hand-operated water sprayer (for ornamental plants) was used to increase the visibility of spider webs. After a spider was encountered, we measured the angle of the web in relation to the ground using a digital angle ruler. We measured web angles because web angles are known to affect rates of prey capture [43]. Similarly, web angles are correlated to wind conditions [44,45]. Moreover, webs are related to habitat with horizontal webs occurring in forest edges and riverine microhabitats [46,47]. We also recorded the number of contact points between web and plants (a proxy of vegetation structure/complexity) [48,49], a rough sketch of the web, plant species the web was constructed on, number of prey items within the web (proxy for successful predation), and GPS coordinates. Spiders were collected into 80 mL plastic vials, one per tube to prevent post-capture predation or cannibalism. After initial collection, specimens were labeled and frozen at $-40\ ^{\circ}$C (to preserve coloration for ease of identification) until further analysis. Once frozen, specimens were sorted and identified by placing them on Petri dishes and then examining them under a dissecting microscope (AmScope, Irvine, CA, USA). Spiders were identified following Ubick et al. [50] and Bradley [51]. In instances where we could not adequately link spiders to species, we identified them by genera. Once specimens were identified and confirmed by a second researcher, ethanol was added to the vials for long-term preservation [51].

2.3. Data Analyses

The effects of vegetation (contact points), prey availability, and web orientation were investigated using a Canonical correspondence analysis (CCA) with abundance data. This technique was particularly appropriate to our data because it addresses the double-zero problem which characterizes community compositional data [52] and does not try to display all variation in the data but only the part that can be explained by the constraints considered [53]. Permutation tests (999 permutations) were run to assess model significance. The sum of the canonical eigenvalues was used as a measure of the variability in the response variables explained by predictors. Analyses were conducted in R using the "vegan" package [54].

We used indicator species analysis IndVal [55] to: (1) identify spider species associated with different parks and (2) identify spider species associated with particular tree species. We chose to use an IndVal analysis because this method identifies indicator species based on specificity (proportion of relative abundances of a taxon found in one park versus other parks) and fidelity (frequency of particular spiders in a particular park). Both parameters range from 0 to 1; specificity is 1 if a taxon is exclusively present in the target treatment, and fidelity is 1 when a taxon is present in all samples of the target treatment. IndVal for each taxon is the product of specificity and fidelity [56]. We performed the analysis at the genus level, used 999 random permutations, and considered spider taxa to be associated with particular parks or tree species if they were significant according to the IndVal permutation tests.

We performed all statistical analyses and graphical representations in R (version 4.1.3, Vienna, Austria) [57]

3. Results

3.1. Species Occurrence and Predictors of Their Occurrence (Question 1)

From the five parks sampled, we recorded 26 species from over 200 individual adults present on webs during the daytime. The most ubiquitous species was the orchard spider (*Leucauge venusta*) followed by basilica orb-weavers (*Mecynogea lemniscata*).

CCA ordination of pooled samples (Figure 2) indicated that variation in spider occurrences was significantly related to prey (correlation = 0.51) and angle of webs (correlation = 0.73). Specifically, high abundances of *Tetragnatha* sp., *Micrathena gracilis*, and *Leucauge venusta* were positively related to prey availability. Web angle was positively correlated to *Mangora maculata* and negatively correlated to *Araneus cavaticus*, *Agelenopsis* sp., and *Argiope aurantia*. Sites with high abundances of spiders such as *Eustala anastera*, *Verrucosa arenata*, *Gea heptagon*, *Atypus affinis*, and *Acanthepeira stellata* were not related to any of the three variables measured (Figure 2). Contact points were not significantly related to spider occurrences.

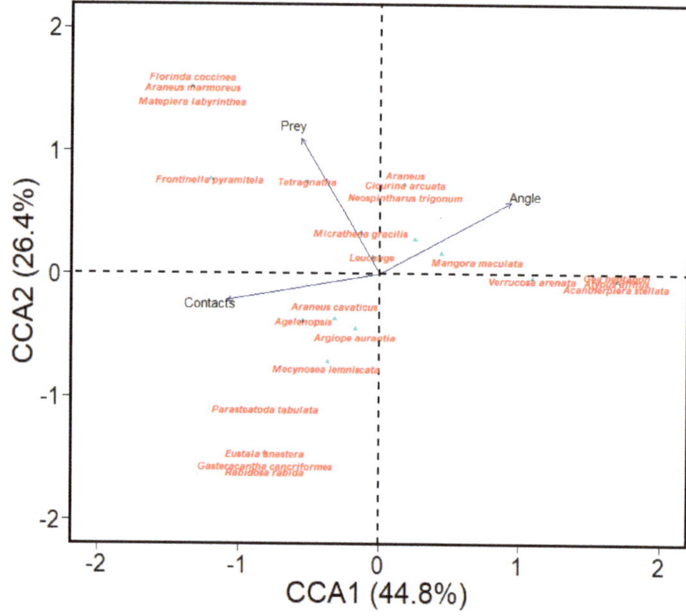

Figure 2. Canonical correspondence analysis (CCA) biplot showing the relationship between spiders and environmental variables (number of contact points (branches), prey abundance enumerated from webs and angle of webs). Samples were collected from parks in Memphis, TN in the summer months of 2022. Blue triangles denote species abundances.

3.2. Habitat Association and Indicator Species (Question 2)

IndVal analysis by habitat and tree species samples revealed the different species that are associated with certain habitats. For instance, Eustala anastera showed an association with Shelby Farms. It is worth noting that the greenlegged orb-weaver (Mangora maculata) was only found from samplings in Nesbit Park (Table 2). Considering the association between spiders and tree species (Table 3) revealed that deadwood was associated with non-web-building spiders (e.g., *Rabidosa rabida*), and web-building spiders (e.g., *Leucauge venusta*) were associated with deadwood in all parks, suggesting the importance of deadwood as a suitable habitat for spiders. Interestingly, *Cicurina arcuata* was only associated with tulip poplar trees. The invasive species Chinese privet (*Ligustrum sinense*) was associated with *Eustala anastera* across all sampled sites.

Table 2. IndVal analysis of species that are associated with specific parks based on the specimens we collected in the summer of 2022. Only significant ($p < 0.05$) habitat associations are presented. The significance of IndVal indices was assessed using 10,000 Monte Carlo permutations.

Park	Species	IndVal	p	Freq
Shelby Farms	*Eustala anastera*	1.00	0.013	3
Shelby Farms	*Parasteatoda tabulata*	1.00	0.01	3
Shelby Farms	*Gasteracantha cancriformes*	1.00	0.014	3
Shelby Farms	*Rabidosa rabida*	1.00	0.009	3
Shelby Farms	*Mecynogea lemniscata*	0.58	0.012	9
Shelby Farms	*Agelenopsis* sp.	0.53	0.012	12
Overton Park	*Cicurina arcuata*	1.00	0.012	3

Table 2. *Cont.*

Park	Species	IndVal	*p*	Freq
Overton Park	*Araneid* sp.	1.00	0.013	3
Overton Park	*Neospintharus trigonum*	1.00	0.016	3
Overton Park	*Leucauge venusta*	0.36	0.014	15
Meeman-Shelby	*Florinda coccinea*	1.00	0.013	3
Meeman-Shelby	*Araneus marmoreus*	1.00	0.009	3
Meeman-Shelby	*Melpomene* sp.	1.00	0.011	3
Meeman-Shelby	*Metepiera labyrinthea*	1.00	0.022	3
Meeman-Shelby	*Frontinella pyramitela*	0.75	0.013	6
T.O Fuller	*Atypus affinis*	1.00	0.01	3
T.O Fuller	*Acantherpiera stellata*	1.00	0.009	3
T.O Fuller	*Gea heptagon*	1.00	0.015	3
T.O Fuller	*Verrucosa arenata*	0.74	0.017	12
Nesbit	*Mangora maculata*	0.50	0.015	15

Table 3. IndVal analysis of species that were associated with specific tree species. Specimens were collected in the summer of 2022. Only significant ($p < 0.05$) habitat associations are presented. The significance of IndVal indices was assessed using 10,000 Monte Carlo permutations.

Tree Species	Species	IndVal	*p*	Freq
Beech	*Araneus marmoreus*	0.83	0.001	5
Beech	*Florinda coccinea*	0.43	0.002	9
Black walnut	*Frontinella pyramitela*	0.33	0.001	36
Black walnut	*Argiope aurantia*	0.27	0.001	61
Black walnut	*Mangora maculata*	0.08	0.001	122
Black walnut	*Mecynogea lemniscata*	0.07	0.001	151
Boxelder	*Micrathena gracilis*	0.11	0.001	120
Boxelder	*Verrucosa arenata*	0.09	0.001	121
Deadwood	*Parasteatoda tabulata*	0.74	0.001	17
Deadwood	*Rabidosa rabida*	0.74	0.001	17
Deadwood	*Agelenopsis pennsylvanica*	0.17	0.016	45
Deadwood	*Leucauge venusta*	0.06	0.001	172
Green ash	*Metepeira labyrinthea*	1.00	0.001	3
Japanese honeysuckle	*Agelenopsis aperta*	0.51	0.001	35
Jumpseed and hornwort	*Leucage* sp.	0.14	0.001	91
Pawpaw	*Gea heptagon*	1.00	0.001	17
Pawpaw	*Araneus cavaticus*	0.71	0.001	23
Pawpaw	*Micrathena sagittata*	0.34	0.001	32
Chinese privet	*Eustala anastera*	0.97	0.001	31
Slippery elm	*Acantherpeira stellata*	1.00	0.001	12
Slippery elm	*Gasteracantha cancriformis*	1.00	0.001	12
Slippery elm	*Tetragnatha* sp.	0.53	0.001	20
Tulip poplar	*Cicurina arcuata*	1.00	0.001	5
Virginia creeper	*Melpomene* sp.	1.00	0.001	5
White oak	*Araneus* sp.	0.60	0.001	9
White oak	*Atypus affinis*	0.40	0.009	6

4. Discussion

We investigated the association of spiders and trees found in parks within a temperate deciduous forest biome. Our results revealed that species such as *Leucauge venusta* and *Mecynogea lemniscata* were the most abundant species recorded across all five parks. One key finding of this work is that the local invasive species Chinese privet is associated with *Eustala anastera*. These findings may have implications for the web-building spider *Eustala anastera* considering that many of these parks have goals to clear all invasives in their parks (Overton Park Conservancy, *personal communication*). Interestingly, other *Eustala* spider species are noted to have associations with particular plants. For instance, researchers in Panama found that *Eustala oblonga* and *Eustala illicita* spiders showed obligate preferences for acacia plants across all studied habitats [58]. Considering this preference for one plant type in many *Eustala* species [58,59], it is tenable that *Eustala anastera* spiders (such as the ones we sampled) may similarly form associations with privet in their environment. Our future work will incorporate greenhouse experiments with different plants to observe potential specificity of *Eustala anastera*.

Our results indicated that differences in plant distribution and spider micro-habitats across different parks (Table 2) are the likely causes of differences in web-building spider communities across these areas. These results confirm a previous study demonstrating that location and habitat characteristics, such as within-field location and landscape features, influence spider presence [60]. For example, spider site preference (tenure), an example of location and landscape features influencing spider presence, has been observed in some long-jawed spiders (*Tetragnatha elongata*), whereby many of member of this species will build webs very close to rivers and only at particular locations [61].

We found that many spider species were associated with particular plants (Table 3) in accordance with other researchers [31,62–64] The association between certain spiders with certain plants across different sites in our study indicates that invasive plant species might interact with specific local spider species in a complex way. Many studies focus on how invasive plants negatively impact herbivorous or native plants [65], and studies on the dynamics between invertebrates and invasive plants offer a more holistic view. Though invasive plants might outcompete native plant species, other native species within ecosystems, such as native invertebrates, can show abilities to adapt to new flora and even benefit from it. For example, researchers have noted the positive benefits that invasive plants can have on spider ecology [16,17]. At the same time, spiders are not the only invertebrates who benefit from the increase in vegetation cover: crickets, slugs, millipedes, and beetle populations are shown to have a positive relationship with vegetation cover too [66].

Our finding further supports the need for case-by-case analysis when resolving the question of invasive species control. In some situations, invasive species eradication has had great success in protecting endangered species [67], and in other situations invasive species can provide wildlife habitat and food resources for local individuals. In the case of Chinese privet, it was introduced to the US in the 1800s, allowing the species to fully expand geographically and co-evolve alongside native species through environmental changes. For invasions at such a large scale, it is time-consuming and economically costly to eradicate the species, and full removal might not return the ecosystem to what it was before the invasion [68]. This is not to say that there should be no control of invasive species or no preservation or protection of native species. In fact, there are certainly native species that do not benefit from invasive plants. For example, many native species in South Africa have been harmed due to the introduction of the invasive Australian *Acacia* causing significant declines in native species richness [69]. However, even with our best efforts, we cannot erase the contribution of globalization and humanity to speeding up the rate of invasive introductions. Scientists should be hesitant to outright list invasive species as automatically bad, and fully examine the impacts invasives have on ecosystems on a case-by-case, species-by-species basis, especially because certain native species have benefitted from invasive species. There is no guarantee that sudden removal of invasives would provide the best benefit to an ecosystem. Due to the benefits invasive plants provide invertebrates through additional surface-area coverage, other studies have suggested seeing whether its beneficial to partially manage invasives, just enough to preserve the benefits that invasives offer certain native species [17], an aspect encapsulated in functional eradication [70]. Thus, further research on how spiders and other invertebrates that inhabit certain invasive plants adjust to their removals is of interest.

Future directions include expanding on this study and consideration of other variables. For example, our study was limited to a specific geographical and temporal range, and as a result, we might not have found certain spider species in certain parks simply because we did not conduct long-term monitoring. Similarly, many spiders build webs at night, but our sampling took place during the daytime, so we might not have found certain spider species because our research had a diurnal spider bias. Having larger sample sizes, conducting more long-term research, conducting research during both night and day, or looking into species-specific analyses on how invasive species impact native species, for example, might give a more holistic and broader view on how invasives impact ecosystem dynamics and give more precise insights into populations in specific areas. Furthermore, expanding research to invertebrates and other relatively understudied species in comparison to mammals offers greater overall insight into how changing plant landscapes are impacting native ecosystems. In terms of spider ecology, future directions might include looking into how spiders in different regions utilize invasive plant species. Our study was limited to five local parks in Southwestern Tennessee; thus, making inferences for other parks in other regions must be conducted with caution. Further studies should investigate how plantation distribution affects prey abundance in answering the association between certain spider species with plant species.

5. Conclusions

In summary, we found that Chinese privet (*Ligustrum sinense*) is associated with the humpbacked orb-weaver (*Eustala anastera*). These findings may have implications for this spider species considering many of these parks have goals to clear all invasives in their parks. Our study makes an effort to contribute to spider research and conservation initiatives in that it provides baseline regional data that can be used to assess the effects of invasive plants in local protected areas. Future studies should consider sampling across larger spatial scales to determine whether the patterns documented here would be true for other ecosystems.

Author Contributions: E.E. and Y.L. have joint first authorship and contributed equally to the manuscript. Methodology, investigation, writing—original draft preparation, E.E. and Y.L.; Conceptualization, writing—reviewing and editing, L.D.C.; Methodology, investigation and editing, A.K.; Supervision, project administration, funding acquisition, writing—reviewing and editing, conceptualization, S.M. All authors have read and agreed to the published version of the manuscript.

Funding: This research was funded by the Office of Fellowship and Undergraduate Research at Rhodes College and Startup Funds granted to S.M. The APC was funded by Rhodes College, Department of Biology.

Data Availability Statement: The datasets analyzed for this study are all contained with the manuscript.

Acknowledgments: We thank B. Jeffcoat for help with field work. We also thank M. Weilmuenster for assistance with sample processing and storage.

Conflicts of Interest: The author declares no conflict of interest. The funders (Rhodes College) and parks sampled had no role in the design of the study; in the collection, analyses, or interpretation of data; in the writing of the manuscript; or in the decision to publish the results.

References

1. Nentwig, W.; Bacher, S.; Kumschick, S.; Pyšek, P.; Vilà, M. More than "100 worst" alien species in Europe. *Biol. Invasions* **2017**, *20*, 1611–1621. [CrossRef]
2. Duncan, C.A.; Jachetta, J.J.; Brown, M.L.; Carrithers, V.F.; Clark, J.K.; Ditomaso, J.M.; Lym, R.G.; McDaniel, K.C.; Renz, M.J.; Rice, P.M. Assessing the Economic, Environmental, and Societal Losses from Invasive Plants on Rangeland and Wildlands. *Weed Technol.* **2004**, *18*, 1411–1416. [CrossRef]
3. Brunson, M.W.; Tanaka, J. Economic and Social Impacts of Wildfires and Invasive Plants in American Deserts: Lessons From the Great Basin. *Rangel. Ecol. Manag.* **2011**, *64*, 463–470. [CrossRef]
4. Bellard, C.; Cassey, P.; Blackburn, T.M. Alien species as a driver of recent extinctions. *Biol. Lett.* **2016**, *12*, 20150623. [CrossRef] [PubMed]
5. Simberloff, D.; Martin, J.-L.; Genovesi, P.; Maris, V.; Wardle, D.A.; Aronson, J.; Courchamp, F.; Galil, B.; García-Berthou, E.; Pascal, M.; et al. Impacts of biological invasions: What's what and the way forward. *Trends Ecol. Evol.* **2013**, *28*, 58–66. [CrossRef]
6. Hoffmann, B.D.; Broadhurst, L.M. The economic cost of managing invasive species in Australia. *NeoBiota* **2016**, *31*, 1–18. [CrossRef]

7. Seebens, H.; Blackburn, T.M.; Dyer, E.E.; Genovesi, P.; Hulme, P.E.; Jeschke, J.M.; Pagad, S.; Pyšek, P.; Winter, M.; Arianoutsou, M.; et al. No saturation in the accumulation of alien species worldwide. *Nat. Commun.* **2017**, *8*, 14435. [CrossRef]

8. Hulme, P.E. Trade, transport and trouble: Managing invasive species pathways in an era of globalization. *J. Appl. Ecol.* **2009**, *46*, 10–18. [CrossRef]

9. Elton, C.S. *The Ecology of Invasions by Animals and Plants*, 2nd ed.; Springer Nature: Cham, Switzerland, 2020; ISBN 978-3-030-34720-8.

10. Doody, J.S.; Green, B.; Rhind, D.; Castellano, C.M.; Sims, R.; Robinson, T. Population-level declines in Australian predators caused by an invasive species. *Anim. Conserv.* **2009**, *12*, 46–53. [CrossRef]

11. Nsengimana, V.; Kaplin, A.B.; Nsabimana, D.; Dekoninck, W.; Francis, F. Diversity and abundance of soil-litter arthropods and their relationships with soil physicochemical properties under different land uses in Rwanda. *Biodiversity* **2021**, *22*, 41–51. [CrossRef]

12. Diagne, C.; Leroy, B.; Vaissière, A.-C.; Gozlan, R.E.; Roiz, D.; Jarić, I.; Salles, J.-M.; Bradshaw, C.J.A.; Courchamp, F. High and rising economic costs of biological invasions worldwide | Nature. *Nature* **2021**, *592*, 571–576. [CrossRef] [PubMed]

13. Crowley, S.L.; Hinchliffe, S.; McDonald, R.A. Conflict in invasive species management. *Front. Ecol. Evol.* **2017**, *15*, 133–141. [CrossRef]

14. Cordell, S.; Ostertag, R.; Michaud, J.; Warman, L. Quandaries of a decade-long restoration experiment trying to reduce invasive species: Beat them, join them, give up, or start over? *Restor. Ecol.* **2016**, *24*, 139–144. [CrossRef]

15. Gleditsch, J.M.; Carlo, T.A. Fruit quantity of invasive shrubs predicts the abundance of common native avian frugivores in central Pennsylvania. *Divers. Distrib.* **2010**, *17*, 244–253. [CrossRef]

16. Landsman, A.P.; Schmit, J.P.; Matthews, E.R. Invasive Plants Differentially Impact Forest Invertebrates, Providing Taxon-Specific Benefits by Enhancing Structural Complexity. *Front. Ecol. Evol.* **2021**, *9*, 682140. [CrossRef]

17. Loomis, J.D.; Cameron, G.N.; Uetz, G.W. Impact of the Invasive Shrub Lonicera maackii on Shrub-Dwelling Araneae in a Deciduous Forest in Eastern North America. *Am. Midl. Nat.* **2014**, *171*, 204–218. [CrossRef]

18. Langellotto, G.A.; Denno, R.F. Responses of Invertebrate Natural Enemies to Complex-Structured Habitats: A Meta-Analytical Synthesis. Available online: https://link.springer.com/article/10.1007/s00442-004-1497-3 (accessed on 12 October 2022).

19. Goodenough, A. Are the ecological impacts of alien species misrepresented? A review of the "native good, alien bad" philosophy. *Community Ecol.* **2010**, *11*, 13–21. [CrossRef]

20. Didham, R.K.; Tylianakis, J.M.; Hutchison, M.A.; Ewers, R.M.; Gemmell, N.J. Are invasive species the drivers of ecological change? *Trends Ecol. Evol.* **2005**, *20*, 470–474. [CrossRef]

21. Schirmel, J.; Bundschuh, M.; Entling, M.H.; Kowarik, I.; Buchholz, S. Impacts of invasive plants on resident animals across ecosystems, taxa, and feeding types: A global assessment. *Glob. Change Biol.* **2016**, *22*, 594–603. [CrossRef]

22. Brown, J.H.; Sax, D.F. An Essay on Some Topics Concerning Invasive Species. *Austral Ecol.* **2004**, *29*, 530–536. [CrossRef]

23. Cash, J.; Anderson, C.; Gulsby, W. The ecological effects of Chinese privet (*Ligustrum sinense*) invasion. *Invasive Plant Sci. Manag.* **2020**, *13*, 3–13. [CrossRef]

24. Birkhofer, K.; Nyffeler, M. An estimated 400–800 million tons of prey are annually killed by the global spider community. *Sci. Nat.* **2017**, *104*, 30. [CrossRef]

25. Buchholz, S. Ground spider assemblages as indicators for habitat structure in inland sand ecosystems. *Biodivers. Conserv.* **2010**, *19*, 2565–2595. [CrossRef]

26. Ameline, C.; Puzin, C.; Bowden, J.J.; Lambeets, K.; Vernon, P.; Pétillon, J. Habitat specialization and climate affect arthropod fitness: A comparison of generalist vs. specialist spider species in Arctic and temperate biomes. *Biol. J. Linn. Soc.* **2017**, *121*, 592–599. [CrossRef]

27. Buchholz, S.; Tietze, H.; Kowarik, I.; Schirmel, J. Effects of a Major Tree Invader on Urban Woodland Arthropods. *PLoS ONE* **2015**, *10*, e0137723. [CrossRef]

28. Harry, I.; Höfer, H.; Schielzeth, H.; Assmann, T. Protected habitats of Natura 2000 do not coincide with important diversity hotspots of arthropods in mountain grasslands. *Insect Conserv. Divers.* **2019**, *12*, 329–338. [CrossRef]

29. Clausen, I.H.S. The use of spiders Araneaeas ecological indicators*. *Br. Arachnol. Soc.* **1986**, *7*, 83–86.

30. Hore, U.; Uniyal, V.; Nentwig, W.; Entling, M.; Kropf, C. Influence of space, vegetation structure, and microclimate on spider (Araneae) species composition in Terai Conservation Area, India. *Eur. Arachnol.* **2010**, *2008*, 71–77.

31. Vasconcellos-Neto, J.; Romero, G.Q.; Santos, A.J.; Dippenaar-Schoeman, A.S. Associations of Spiders of the Genus Peucetia (Oxyopidae) with Plants Bearing Glandular Hairs. *Biotropica* **2007**, *39*, 221–226. [CrossRef]

32. Scheidler, M. Influence of habitat structure and vegetation architecture on spiders. *Zool. Anz.* **1990**, *225*, 333–340.

33. Knight, D.B.; Davis, R.E. Climatology of Tropical Cyclone Rainfall in the Southeastern United States. *Phys. Geogr.* **2007**, *28*, 126–147. [CrossRef]

34. Schmid, P.E. Observing and Modeling Urban Thunderstorm Modification Due to Land Surface and Aerosol Effects. Ph.D. Thesis, Purdue University Graduate School, West Lafayette, IN, USA, 2020. [CrossRef]

35. Kahle, D.; Wickham, H. ggmap: Spatial Visualization with ggplot2. *R J.* **2013**, *5*, 144–161. [CrossRef]

36. Meeman-Shelby Forest State Park Meeman-Shelby Forest State Park Business & Management Plan 2018. Available online: https://tnstateparks.com/assets/pdf/park-plans/June2018MeemanShelbyBP.pdf (accessed on 24 August 2022).

37. West, C.V. Meeman-Shelby Forest State Park. Available online: https://tennesseeencyclopedia.net/entries/meeman-shelby-forest-state-park (accessed on 3 October 2022).

38. Trust for Public Land Tennessee Office The Parks of Memphis—Past, Present & Future 2014. Available online: https://www.tpl.org/wp-content/uploads/2015/07/ccpe-memphis-parks-report.pdf (accessed on 10 September 2022).

39. Green, J. Sampling Method and Time Determines Composition of Spider Collections. *J. Arachnol.* **1999**, *27*, 176–182.

40. Ausden, M.; Drake, M. Invertebrates. In *Ecological Census Techniques*; Sutherland, W., Ed.; Cambridge University Press: Cambridge, UK, 2006; pp. 214–249.

41. McCaffrey, J.P.; Parrella, M.P.; Horsburgh, R.L. Evaluation of the Limb-Beating Sampling Method for Estimating Spider (Araneae) Populations on Apple Trees. *J. Arachnol.* **1983**, *11*, 363–368.

42. Whitmore, C.; Slotow, R.; Crouch, T.E.; Dippenaar-Schoeman, A.S. Diversity of spiders (Araneae) in a Savanna reserve, Northern province, South Africa. *Arac* **2002**, *30*, 344–356. [CrossRef]

43. Bishop, L.; Connolly, S.R. Web Orientation, Thermoregulation, and Prey Capture Efficiency in a Tropical Forest Spider. *J. Arachnol.* **1992**, *20*, 173–178.

44. Eberhard, W.G. The ecology of the web of *Uloborus diversus* (Araneae: Uloboridae). *Oecologia* **1971**, *6*, 328–342. [CrossRef]

45. Wu, C.-C.; Blamires, S.J.; Wu, C.-L.; Tso, I.-M. Wind induces variations in spider web geometry and sticky spiral droplet volume. *J. Exp. Biol.* **2013**, *216*, 3342–3349. [CrossRef]

46. Buskirk, R.E. Coloniality, Activity Patterns and Feeding in a Tropical Orb-Weaving Spider. *Ecology* **1975**, *56*, 1314–1328. [CrossRef]

47. Eberhard, W.G. Function and Phylogeny of Spider Webs. *Annu. Rev. Ecol. Syst.* **1990**, *21*, 341–372. [CrossRef]

48. Balfour, R.A.; Rypstra, A.L. The Influence of Habitat Structure on Spider Density in a No-Till Soybean Agroecosystem. *J. Arachnol.* **1998**, *26*, 221–226.

49. Mcnett, B.J.; Rypstra, A.L. Habitat selection in a large orb-weaving spider: Vegetational complexity determines site selection and distribution. *Ecol. Entomol.* **2000**, *25*, 423–432. [CrossRef]

50. Ubick, D.; Cushing, P.E. *Spiders of North America: An Identification Manual, Second Edition*, 2nd ed.; American Arachnological Society: Keene, NH, USA, 2017; ISBN 978-0-9980146-0-9.

51. Bradley, R.A. *Common Spiders of North America*, 1st ed.; University of California Press: Berkeley, CA, USA, 2019; ISBN 978-0-520-31531-0.

52. Legendre, P.; Gallagher, E.D. Ecologically meaningful transformations for ordination of species data. *Oecologia* **2001**, *129*, 271–280. [CrossRef]

53. Oksanen, J.; Blanchet, F.G.; Kindt, R.; Legendre, P.; O'Hara, R.; Simpson, G.; Solymos, P.; Stevens, M.; Wagner, H. Package 'vegan': Community Ecology Package. Version 1.17–2. 2010. Available online: https://www.researchgate.net/publication/282247921_Package_%27vegan%27_Community_Ecology_Package_Version_117--2 (accessed on 22 November 2022).

54. Oksanen, J.; Blanchet, F.G.; Kindt, R.; Legendre, P.; Minchin, P.; O'Hara, B.; Simpson, G.; Solymos, P.; Stevens, H.; Wagner, H. Vegan: Community Ecology Package. R Package Version 2.2–1. 2015, Volume 2, pp. 1–2. Available online: http://www.pelagicos.net/MARS6910_spring2015/manuals/R_vegan.pdf (accessed on 22 November 2022).

55. Dufrêne, M.; Legendre, P. Species Assemblages and Indicator Species:the Need for a Flexible Asymmetrical Approach. *Ecol. Monogr.* **1997**, *67*, 345–366. [CrossRef]

56. Wildi, O.; Feldmeyer-Christe, E. Indicator values (IndVal) mimic ranking by F-ratio in real-world vegetation data. *Community Ecol.* **2013**, *14*, 139–143. [CrossRef]

57. R Core Team. *A Language and Environment for Statistical Computing*; R Foundation for Statistical Computing: Vienna, Austria, 2022.

58. Ledin, A.E.; Styrsky, J.D.; Styrsky, J.N. Friend or Foe? Orb-Weaver Spiders Inhabiting Ant–Acacias Capture Both Herbivorous Insects and Acacia Ant Alates. *J. Insect Sci.* **2020**, *20*, 16. [CrossRef]

59. Hesselberg, T.; Triana, E. The web of the acacia orb-spider Eustala illicita (Araneae: Araneidae) with notes on its natural history. *J. Arachnol.* **2010**, *38*, 21–26. [CrossRef]

60. Clough, Y.; Kruess, A.; Kleijn, D.; Tscharntke, T. Spider diversity in cereal fields: Comparing factors at local, landscape and regional scales. *J. Biogeogr.* **2005**, *32*, 2007–2014. [CrossRef]

61. Smallwood, P.D. Web-Site Tenure in the Long-Jawed Spider: Is It Risk-Sensitive Foraging, or Conspecific Interactions? *Ecology* **1993**, *74*, 1826–1835. [CrossRef]

62. Whitney, K.D. Experimental Evidence That Both Parties Benefit in a Facultative Plant-Spider Mutualism. *Ecology* **2004**, *85*, 1642–1650. [CrossRef]

63. Dennis, P.; Young, M.R.; Gordon, I.J. Distribution and abundance of small insects and arachnids in relation to structural heterogeneity of grazed, indigenous grasslands. *Ecol. Entomol.* **1998**, *23*, 253–264. [CrossRef]

64. Malumbres-Olarte, J.; Vink, C.J.; Ross, J.G.; Cruickshank, R.H.; Paterson, A.M. The role of habitat complexity on spider communities in native alpine grasslands of New Zealand. *Insect Conserv. Divers.* **2013**, *6*, 124–134. [CrossRef]

65. Sieg, C.H.; Phillips, B.G.; Moser, L.P. Exotic invasive plants. In *Ecological Restoration of Southwestern Ponderosa Pine Forests*; Frederici, P., Ed.; Island Press: Washington, DC, USA, 2003; pp. 251–267.

66. Norbury, G.; Heyward, R.; Parkes, J. Skink and invertebrate abundance in relation to vegetation, rabbits and predators in a New Zealand dryland ecosystem on JSTOR. *N. Z. J. Ecol.* **2009**, *33*, 24–31.

67. Hayes, S.J.; Holzmueller, E.J. Relationship between Invasive Plant Species and Forest Fauna in Eastern North America. *Forests* **2012**, *3*, 840–852. [CrossRef]

68. Wang, H.-H.; Grant, W.E. Invasion of Eastern Texas Forestlands by Chinese Privet: Efficacy of Alternative Management Strategies. *Diversity* **2014**, *6*, 652–664. [CrossRef]

69. Gaertner, M.; Breeyen, A.D.; Hui, C.; Richardson, D.M. Impacts of alien plant invasions on species richness in Mediterranean-type ecosystems: A meta-analysis. *Prog. Phys. Geogr. Earth Environ.* **2009**, *33*, 319–338. [CrossRef]

70. Green, S.J.; Grosholz, E.D. Functional eradication as a framework for invasive species control. *Front. Ecol. Environ.* **2021**, *19*, 98–107. [CrossRef]

insects

Review

Host Plant Specificity in Web-Building Spiders

Thomas Hesselberg [1,2,*], Kieran M. Boyd [3], John D. Styrsky [4] and Dumas Gálvez [5,6,7]

1. Department for Continuing Education, University of Oxford, Oxford OX1 2JA, UK
2. Department of Biology, University of Oxford, Oxford OX1 3SZ, UK
3. School of Biological Sciences, Queen's University Belfast, Belfast BT7 1NN, UK
4. Department of Biology, University of Lynchburg, Lynchburg, VA 24501, USA
5. Coiba Scientific Station, Panama City 0843-01853, Panama
6. Programa Centroamericano de Maestría en Entomología, Universidad de Panamá, Panama City 0824, Panama
7. Smithsonian Tropical Research Institute, Panama City P.O. Box 0843-03092, Panama
* Correspondence: thomas.hesselberg@conted.ox.ac.uk

Simple Summary: Many invertebrates interact and are associated with plants in nature. However, despite their abundance and ecological importance, our knowledge of spiders and their associations with plants is limited. Here, we review what we currently know about spider–plant interactions and associations, with a focus on web-building spiders. This includes an overview of the most prominent interactions non-web-building and web-building spiders have with plants, followed by examples of the specific web-building spider–plant associations we know of, where especially the Acacia–*Eustala* association observed in Panama is interesting. We also review the plausible mechanisms for host plant location and finally present some ideas for future research.

Abstract: Spiders are ubiquitous generalist predators playing an important role in regulating insect populations in many ecosystems. Traditionally they have not been thought to have strong influences on, or interactions with plants. However, this is slowly changing as several species of cursorial spiders have been reported engaging in either herbivory or inhabiting only one, or a handful of related plant species. In this review paper, we focus on web-building spiders on which very little information is available. We only find well-documented evidence from studies of host plant specificity in orb spiders in the genus *Eustala*, which are associated with specific species of swollen thorn acacias. We review what little is known of this group in the context of spider–plant interactions generally, and focus on how these interactions are established and maintained while providing suggestions on how spiders may locate and identify specific species of plants. Finally, we suggest ideas for future fruitful research aimed at understanding how web-building spiders find and utilise specific plant hosts.

Keywords: spider–plant interactions; swollen thorn acacias; carnivorous plants; orb-web spiders; host recognition; plant volatiles

Citation: Hesselberg, T.; Boyd, K.M.; Styrsky, J.D.; Gálvez, D. Host Plant Specificity in Web-Building Spiders. *Insects* **2023**, *14*, 229. https://doi.org/10.3390/insects14030229

Academic Editor: Cheryl Y. Hayashi

Received: 3 January 2023
Revised: 3 February 2023
Accepted: 21 February 2023
Published: 24 February 2023

1. Introduction

Plants are a vital resource for many animals that use them for food, shelter or protection. The best known plant–animal interactions involve insects and include negative interactions, such as herbivory, and positive interactions, such as pollination, and other mutualistic interactions. In many of these interactions, the insect shows specificity in that it only interacts with one, or a couple of plant species. These examples can be tightly co-evolved and include the food-for-protection mutualism between ants and swollen thorn acacias, where a specific species of ant is paired with a specific species of acacia [1], the extreme specificity of fig wasp pollinators to particular fig species hosts [2], and the specialisation of small groups of orchids to one species of bee pollinator, such as the South African guild of orchids (Coryciinae) exclusively relying on the oil-collecting bee (*Rediviva peringueyi*) for pollination [3].

Insects, as outlined above, and other arthropods, such as herbivorous and mutualistic mites [4], are well known for developing close associations with plants. Spiders, however, are usually thought of as generalist predators that only use vegetation indiscriminately for shelter or as a substrate for their webs. A study on a temperate grassland spider community, for example, showed that while some individual spider species showed a weak preference for a narrow range of host plants, the overwhelming preference was for tall and stable vegetation structures and not individual plant species [5]. Recently, the long-held notion that spiders have limited interactions with the vegetation in their surroundings have been challenged, especially by the surprising discovery that some species of spiders, and the first instars of web-building spiders in particular, rely on nectar, pollen and Beltian bodies as a significant component of their diets [6–8]. This prompted a review of spider–plant interactions in general, which revealed associations with plants across a much larger range of spider families than previously thought [9].

Very limited research is available on spiders that construct aerial webs, which predominantly consist of sheet-webs by members of the family Linyphiidae, tangle webs by members of the family Theriididae, and orb webs by members of the families Araneidae and Tetragnathidae. As the function of the webs to some degree depends on the substrate to which they are attached, it could be argued that they are more dependent on the correct choice of plant, and therefore, potentially should be more discerning than cursorial spiders. A relatively newly described species of linyphiid, *Laetesia raveni*, from Australia appears to exclusively build its webs on two thorny plant species, *Calamus muelleri* and *Solanum inaequilaterum* [10]. Similarly, one genus of araneid spiders, *Eustala*, seems a promising candidate for more in-depth research as several studies show close associations to individual species of acacias in the genus *Vachellia* [11,12]. These acacia species are in a mutualistic relationship with protective *Pseudomyrmex* ants, and the *Eustala* spiders probably associate closely with the acacias to exploit the ant–acacia mutualism for enemy-free space [13].

Another largely unresolved question is how spiders locate and identify their host plants. Insects generally locate their host plants using chemical cues from wind-dispersed plant volatiles [14,15]. In ant–plant associations, ants identify their mutualistic partner by chemical cues emitted from the plant [16]. However, the distance to which they rely on plant volatiles, or random searches for the location of host plants remains unclear. On the one hand, *Pheidole minutula* used plant volatiles to correctly locate their host plant *Maieta guianensis* during choice tests over distances of 15 cm in Y-maze experiments in the laboratory [17], while on the other hand, *Crematogaster* ants recognise their host *Macaranga* species only by direct contact with chemical compounds on the stem surface of saplings [18]. Spiders are also known to use chemical cues during mating behaviour [19], such as males using cues from silk to locate and evaluate females [20], and they use them to detect potential prey [21]. In addition, there are a few examples of spiders using chemical cues from plants, including two species of crab spiders in the genus *Thomisus* that were attracted to the clove oil flower fragrance [22] and the nectivorous spider *Hibana futilis*, which uses plant volatiles to recognise and potentially locate nectar sources [6].

The main aim of this review paper is to review the limited data we have on host plant specificity in web-building spiders and to contrast it with what is known from cursorial spiders. Secondarily, we review the limited literature on how web-building spiders identify and find web-building locations, including suitable plants, using the above-mentioned *Eustala* orb spiders as a model system. We hope to stimulate further research by identifying significant gaps and outlining promising experimental approaches to plug some of these gaps.

2. Spider–Plant Associations

In this section, we provide a brief overview of some of the best described examples of close spider–plant associations for both cursorial (i.e., non-web-building) and web-building spiders. This has been recently reviewed by Vasconcellos-Neto et al. [9], but here we update with newer references focussing predominantly on web-building spiders and link the topic to host plant locations in general and the *Eustala*–acacia–ant system in particular.

2.1. Cursorial Spiders

Bromeliads and other rosette-structured plants have a complex, three-dimensional architecture that presents a valuable microhabitat for a number of species [23], particularly members of Salticidae [24–26]. The best studied cursorial spider–plant association, and one of the few species-specific examples, is that of the bromeliad specialist *Psecas chapoda* and *Bromelia balansae*. Through a series of studies by Romero and Vasconcellos-Neto [27–29], *P. chapoda* was found exclusively on *B. balansae* across a large geographic range [26] (Table 1). Whilst *B. balansae* provides *P. chapoda* with a favourable microhabitat and microclimate, *P. chapoda* has been reported to contribute to the nutrition of *B. balansae* through the absorption of nitrogen from spider faeces deposited on the leaves of the bromeliad [30]. Romero et al. [31,32] evidenced that this interaction was indeed mutualistic as the leaves of *B. balansae* grew larger in the presence of *P. chapoda*.

Some Thomisidae crab spiders, which have been documented as obligate *Nepenthes* pitcher-plant dwellers (Table 1), have likewise been reported to assist their host plant with nitrogen acquisition. The specialised leaves of pitcher-plants, which are used to attract, trap, and digest prey [33,34], also provide suitable microhabitats for the crab spiders *Misumenops nepenthicola* and *Thomisus nepenthephilus* [34,35]. These spiders feed on visiting insects drawn to the pitcher-plants [34,36], and in some circumstances, the spiders increase pitcher-plant prey consumption by dropping consumed prey remains into the pitchers. Interestingly, two studies by Lim et al. [34], and Lam and Tan [37] concluded that the type of association between crab spiders and pitcher-plants is environmentally context-dependent. Lam and Tan [37] demonstrated that *T. nepenthephilus* increased the prey capture rates of *Nepenthes gracilis*, offsetting the nitrogen loss from consumption by *T. nepenthephilus*, resulting in an overall net gain. However, this benefit only occurs under conditions where prey availability is low and is ultimately lost when prey availability increases, switching from a positively facilitative to a parasitic interaction [37].

Furthermore, a number of spider species have been reported to have unusually close associations with trichome-bearing plants [9,38–41]. One genus from the Oxyopidae family, *Peucetia*, dominates such interactions and many species are considered to have strict, and perhaps obligatory, associations with glandular trichome-bearing plants [39,40,42]. Glandular trichomes are hair-like structures believed to have evolved as a direct biotic defence against herbivorous insects [43,44]. The insects and carrion (i.e., dead insects) trapped by the glandular hairs represent an energetically cost-free, accessible food source [45], which attracts arthropod predators, such as spiders, for added protection against herbivory [40,45,46]. In three complementary studies, Morais-Filho and Romero [39,40,47] observed *Peucetia flava* exclusively in association with *Rhyncanthera dichotoma*. During the latter study, Morais-Filho and Romero [40] physically removed the glandular trichomes from *R. dichotoma* and documented fewer *Peucetia* spiders occupying those plants compared to *R. dichotoma* with intact trichomes, further demonstrating the strong and potentially obligatory association *Peucetia* spiders have with glandular trichome-bearing plants [42]. Morais-Filho and Romero [40] reported that *P. flava* reduced herbivory in the buds and flowers of *R. dichotoma* and although this interaction did not increase fruit production, it also did not incur any significant costs to *R. dichotoma* fitness (i.e., through predation of pollinators), signifying a potential protective mutualism. Moreover, a recent study by Sousa-Lopes et al. [45] found that the presence of *P. flava* on the trichome-bearing *Mimosa setosa* var. *paludosa* positively correlated with an increase in trapped prey and carrion.

Spider–plant associations that arise from an exploitable source of food are not uncommon. While some spiders may associate with plants that attract and/or trap insect prey, such as glandular trichome-bearing plants and pitcher-plants, other spiders species seek nutrition from the plant itself. The salticid, *Bagheera kiplingi*, for example, is exclusively associated with many myrmecophytic acacias [7,48]. These acacias produce Beltian bodies to attract ants that protect the plant, and in return, the ants gain nutritional rewards and refuge [1,7]. The spider exploits this ant–acacia mutualism and consumes the Beltian bodies as its primary food source, which in some cases constitute 90% of its diet [48]. Therefore, it is conceivable that access to a convenient source of prey is another primary driver of spider–plant associations, and perhaps the obligatory associations observed between *Peucetia* and glandular trichome-bearing plants and Thomisidae and *Nepenthes* pitcher-plants.

Another potential driver of host plant selectivity in spiders could be crypsis (i.e., camouflage), whereby a spider may exhibit a preferential affinity for a substrate (e.g., flower, bark, and moss) that matches their body colouration/morphology, rendering them undetectable to potential predators or unsuspecting prey. Cryptic colouration is particularly well studied in Thomisidae crab spiders, which, in sit-and-wait predators, increases foraging success [49–51]. Certain species will preferentially select flowers, upon which they forage, that match their body colouration (i.e., background-matching) to avoid detection by pollinators and other visiting insects [41,49,52]. Moreover, there are some spider species that are also capable of changing their body colouration to match their chosen background, or in this instance, host plant. Such examples include the crab spiders *Misumena vatia* and *Thomisus onustus* that typically alternate between white and yellow [50,53]. It is evident that cryptic species will select specific substrates to ensure successful camouflage. However, there is a paucity of information to discern whether cryptic colouration is a resultant factor in specific spider–plant associations. Most crab spiders appear to be generalists, selecting a number of plant species that suit their needs.

From the examples provided above, it is particularly apparent that *Psecas chapoda* facultatively relies on the microhabitat created by *B. balansae* for foraging, mating, and oviposition, as observed by Romero and Vasconcellos-Neto [28,29], and as a refuge and nursery site that can offer protection from predators and desiccation [28,29,54,55]. Omena and Romero [56] inferred that this extreme fidelity was related to microhabitat structure, and observations by Romero and Vasconcellos-Neto [28,55] affirmed this after finding that *P. chapoda* seldom colonised bromeliads in forest habitats as leaves would often obstruct the rosette, hindering any use of the microhabitat. Likewise, some studies have reported that *Peucetia* spiders preferentially select larger plants as they offer more sites to forage and refuge, and attract and trap more insect prey [45,57]. Prey, and other sources of nutrients, are also key determinants, especially in terms of exploitable sources of food, which we see examples of in all three spider families discussed. In summary, we can infer that it is the availability of certain exploitable resources, together with a microhabitat structure and plant morphology that complements the ecological requirements, foraging the strategies and behavioural preferences of a spider [9,23,56,58–61], which are the primary factors that drive specific spider–plant associations.

Table 1. The most prominent cursorial spider–plant associations. With information on the spider and host plant family and species, information on the association, and the location(s) where said interaction was documented.

Spider Family	Spider Species	Plant Family	Plant Species	Association	Region	Source
Oxyopidae	*Peucetia flava*	Asteraceae	*Trichogoniopsis adenantha*	Facultative mutualism; reduced herbivores.	Southeast Brazil	Romero et al. [46]
		Melastomataceae	*Rhyncanthera dichotoma*	Commensalism/facultative mutualism; protection and significantly reduced herbivory after rainy season.	Southeast Brazil	Morais-Filho and Romero [39,40,47]
		Solanaceae	*Solanum thomasiifolium*	Facultative mutualism; likely protection.	Southeast Brazil	Jacobucci et al. [57]
		Fabaceae	*Mimosa setosa var. paludosa*	Facultative mutualism; reduced exophytic herbivory, but not endophytic herbivory.	Southeast Brazil	Sousa-Lopes et al. [45]

Table 1. *Cont.*

Spider Family	Spider Species	Plant Family	Plant Species	Association	Region	Source
	Peucetia rubrolineata	Asteraceae	*Trichogoniopsis adenantha*	Facultative mutualism; suppressed herbivory.	Southeast Brazil	Romero et al. [46]
	Peucetia viridans	Euphorbia-ceae	*Cnidoscolus aconitifolius*	Preference/Unknown	Southeast Mexico	Arango et al. [62]
			Croton ciliatoglandulifer	Preference/Unknown	West Mexico	Corcuera et al. [63]
	Bagheera kiplingi	Fabaceae	*Vachellia spp.* (myrmecophytes)	Exploitative/Commensalism	Southeast Mexico, Northwest Costa Rica	Meehan et al. [7]
	Evarcha culicivora	Euphorbia-ceae	*Ricinus communis*	Unknown/Commensalism	West Kenya	Cross [64]
			Lantana camara	Unknown/Commensalism	West Kenya	Cross [64]
Salticidae	*Pelegrina tillandsiae*	Bromeliaceae	*Tillandsia usneoides*	Obligate commensalism; strict association, but no reported costs or benefits.	Southeast USA	Young and Lockley [24]
	Psecas chapoda	Bromeliaceae	*Bromelia balansae*	Facultative mutualism; the spider aids in nitrogen acquisition.	Northeast Bolivia, Northeast Paraguay, South Brazil, Central-West Brazil	Romero [26]; Romero and Vasconcellos-Neto [29]; Romero et al. [31]; Omena and Romero [56]
Sparrasidae	*Delena melanochelis*	Myrtaceae	*Eucalyptus nitens*	Unknown/Commensalism	Australia	Agnarsson and Rayor [65]
			E. regnans	Unknown/Commensalism	Australia	Agnarsson and Rayor [65]
Thomisidae	*Misumenops argenteus*	Lamiaceae	*Hyptis suaveolens*	Unknown/Commensalism	Southeast Brazil	Romero and Vasconcellos-Neto [27]
		Asteraceae	*Trichogoniopsis adenantha*	Facultative mutualism; reduced herbivory	Southeast Brazil	Romero and Vasconcellos-Neto [27,55]
	Misumenops pallidus	Orchideaceae	*Chloraea alpina*	Commensalism	East Argentina	Quintero et al. [66]
		Ranunculaceae	*Anemone multifida*	Commensalism	East Argentina	Gavini et al. [11]
	Misumenops nepenthicola	Nepentha-ceae	*Nepenthes gracilis*	Unknown/Commensalism	North Borneo	Karl and Bauer [67]
			N. rafflesiana	Unknown/Commensalism	North Borneo	Karl and Bauer [67]
	Synaema marlothi	Roridulaceae	*Roridula dentata*	Obligate kleptoparasitism	Southern South Africa	Anderson and Midgley [38]; Anderson [68]
	Synaema obscuripes	Nepentha-ceae	*Nepenthes madagascariensis*	Unknown	Southeast Madagascar	Rembold et al. [36]
	Thomisus nepenthephilus	Nepentha-ceae	*Nepenthes gracilis*	Obligate, conditional facilitative mutualism	North Singapore	Lim et al. [34]; Lam and Tan [37]

2.2. Web-Building Spiders

Research on web-building spider–plant associations is far less numerous than on their non-web-building counterparts. Currently, there are only a few examples of exclusive spider–plant associations, represented by *Eustala* (Araneidae) and *Laetesia raveni* (Linyphiidae), which are discussed in more detail in Section 3 below. The research on cursorial spider–plant associations indicates that the suitability of a plant as a microhabitat to find shelter or food resources (i.e., prey, carrion or nectar) are the main determinants of host plant selection and subsequent spider–plant associations. This also applies to web-building species, where it is vital to select a web-building site that maximises foraging success [58]. For these sit-and-wait predators this is ultimately dependent on the density of prey [69,70], which as mentioned is a key driver in host plant selection. However, the key driver of foraging success for a web-building spider is the optimal construction of its web; hence, the majority of available research on web-building spiders documents preferential, facultative associations with plants that provide suitable structural features for web construction [48,71–74].

Two neotropical spider species, the theridiid *Latrodectus geometricus* and the araneid *Alpaida quadrilorata*, are both found in association with *Paepalanthus bromelioides* [73]. This rosette-structured plant provides the spiders with the structural necessities for web construction and may also offer refuge and protection from predators [23,75]. More importantly, *P. bromelioides* is considered to be a protocarnivorous plant that obtains nutrients from insects with the aid of digestive mutualists, namely *L. geometricus* and *A. quadrilorata* [73]. This plant apparently possesses features that attract insect prey, such as leaves that reflect ultraviolet light and a phytotelma (i.e., a water-filled cavity) with specialised fluid that also digests captured prey [73,76]. Similar to the pitcher-plant dwelling Thomisidae crab spiders that forage at the mouth of the pitcher, *L. geometricus* and *A. quadrilorata* build their webs above the phytotelma [76], providing easy access to incoming prey. Both spider species capture prey, while discarding carcasses and faeces into the rosette of *P. bromelioides* effectively, and thereby channelling a more bioavailable form of nitrogen directly to the plant [31]. Nishi et al. [73] observed *A. quadrilorata* strictly on *P. bromelioides* within the study area in Morro da Pedreira, Brazil. However, no other research is available to determine how exclusive this association is, and since *L. geometricus* has been documented on other plant species (e.g., [68]), both should be considered facultative digestive mutualists.

As previously discussed, carnivorous plants present spiders with a suitable microhabitat [34,37]. However, aside from Nishi et al. [73], there are no reports of unequivocal web-building spider associations with carnivorous plants. Cresswell [77] observed an unidentified species of linyphiid occupying the pitcher-plant *Sarracenia purpurea* as an apparent kleptoparasite. Milne and Waller [78] similarly observed linyphiids interacting with *S. purpurea*, using the pitchers as substrates to build their horizontal sheet webs. However, Milne and Waller [78] noted that many of the linyphiids constructed their webs at a height similar to the pitchers, implying that this a spatial coincidence rather than an association. The theridiid *Theridion decaryi* has also been observed inhabiting a different pitcher-plant species, *Nepenthes madagascariensis*, according to Fage [79]. The available research on these interactions is evidently scarce and ambiguous. However, considering that several other spider species have been found in association with pitcher-plants and other carnivorous plants (Table 1), the possibility that there are species of web-building spiders closely associated with pitcher-plants cannot be ruled out.

In addition, web-building spiders in the genus *Stegodyphus* (Eresidae) have strong affinities for thorny plants [54,72,80]. A recent study by Rose et al. [54] determined that *Stegodyphus dumicola* nests occurred more frequently on tall thorny plants and were observed on several different genera. Lubin et al. [80] also found that *S. lineatus* preferred to inhabit tall, thorny, and even poisonous plants. Thorny plants offer protection against predators (e.g., birds) and reduce the risk of disturbances from large herbivorous animals (e.g., cattle and other browsing/grazing mammals) that can damage or destroy spider webs [54,72,75,80]. Ruch et al. [72] demonstrated that *S. tentoriicola*, which inhabits both thorny and thornless plants, constructed larger webs when inhabiting thorny plants, and were less likely to relocate, compared to spiders in thornless vegetation. As larger webs are more costly to build, it is evident that thorny plants provide *S. tentoriicola*, and likely other spider occupants, with favourable microhabitats that enable spiders to invest more energy into building larger webs, increasing their foraging success, whilst receiving refuge and protection from animal-related disturbances [54,72].

Extreme specificity and fidelity toward host plants is evidently not as common among web-building spiders. Many web-building spiders often interact with and inhabit multiple plant species from different families and orders, as described, for example, by Rose et al. [54] and Whitney [71]. A recent study conducted by Cuff et al. [81] in England evaluated the leaf and habitat preferences for oviposition in the candy-striped spiders *Enoplognatha ovata* and *E. latimana* in the family Theridiidae. These spiders create a retreat, or nest, for oviposition by rolling a leaf with silk [81]. *Enoplognatha* appeared to preferentially select the leaves of bramble (*Rubus fruticosus*), nettle (*Urtica dioica*), hogweed (*Heracleum sphondylium*), and have also been found using fireweed (*Chamaenerion angustifolium*) for their leaf-roll nests. Plant preferences were not taxon-related, nor was the size and structure of leaves important; however, certain traits, such as the length–width ratio, were thought to influence leaf selection [81]. Cuff et al. [81] even suggested that the spiders could possibly provide a degree of protection from herbivorous insects in a mutualistic association.

3. Host Plant Specificity in Web-Building Spiders: The Unique Cases of *Eustala* and *Laetesia*

Very few one-to-one obligatory associations between spiders and specific plants have been described and, as mentioned above, most of these involve cursorial spiders. Here, we discuss the two examples we found: one in the orb-web genus *Eustala* (family Araneidae) and one in the sheet-weaver *Laetesia raveni* (family Linyphiidae).

3.1. The Araneid Orb-Web Spiders in the Genus Eustala

Species in the orb-web family Araneidae commonly inhabit plants on which they construct their webs. Although none of the genera in this large family are characterised as being closely associated with particular plant groups or plant species, recent work indicates that several species in the genus *Eustala* exhibit varying degrees of host plant specificity. The genus *Eustala* is large with around 90 species distributed throughout North and South America, the majority of which are found at tropical latitudes [82–84]. Early studies of the natural histories of the *Eustala* species noted that they do not typically build a retreat but rather rest on branches or are tucked into dead vegetation that they resemble in colour and pattern near their webs (e.g., [85,86]).

Eustala perfida, for example, exhibits a colour polymorphism that closely resembles the mosses and lichens on the tree trunks on which it builds its webs in semi-deciduous rainforests in south eastern Brazil. A detailed study of spatial distribution and substrate selection showed that this spider apparently prefers specific microhabitats characterised by large-diameter rough-barked trees with mosses, lichens, and concavities, but that it does not uniquely inhabit the bark of any one particular tree species [87]. Two other *Eustala* species in south eastern Brazil, *E. sagana* and *E. taquara*, however, show a closer association with particular plant species. Both spider species preferentially rest on dead vegetation, against which they are strongly camouflaged, versus live vegetation (see images in Souza et al. [88]). A comparison of the relative frequencies of plant species in the spiders' preferred edge habitats with the relative frequencies of plant species used for web construction provides evidence for some level of host plant specificity. *Eustala taquara* occupied the fleabane *Conyza bonariensis* significantly more frequently than other plant species, whereas *E. sagana* significantly more frequently occupied a different weedy plant species, *Hyptis suaveolens* [88]. Preferential use of these plant species for web construction and retreats may reduce conflict between the two spider species in an area of range overlap along an elevation gradient [88].

The evidence for even stronger associations between *Eustala* spiders and specific plant species comes from research in central Panama. *Eustala oblonga* and *E. illicita* are found in abundance on the ant acacias *Vachellia melanoceras* and *V. collinsii* on the Atlantic and Pacific sides of the Continental Divide, respectively [11,12,89]. Remarkably, on plants on which patrolling acacia ant mutualists tolerate few other animal interlopers, these two spider species construct webs at night and rest by day on the acacia leaves, branches, and thorns, where they are mostly ignored (or undetected) by the ants; they also breed and construct egg sacs on the acacias (Figure 1). Neither *E. oblonga* nor *E. illicita* prey on patrolling ants, but they do capture dispersing acacia ant alates in their webs in addition to many other flying insects [89]. Surveys of 50 *V. melanoceras* acacias, 50 neighbouring non-acacias (J.D. Styrsky and J.N Styrsky, unpublished data), 18 *V. collinsii* acacias, and 18 neighbouring non-acacias [11] showed that both *E. oblonga* and *E. illicita*, respectively, are found almost exclusively on ant acacias (Figure 2). Although neither spider is typically encountered elsewhere in the forest understory, a few individuals of both species were observed resting on dead, weedy, roadside vegetation in Parque Nacional Soberania and Parque Natural Metropolitano, respectively (T. Hesselberg and J. Styrsky, unpublished observations), raising the possibility that their association with ant acacias may not be entirely obligatory.

Figure 1. (**A**) A female (left) and (**B**) male (right) *Eustala oblonga* on the foliage of *Vachellia melanoceras* in Parque Nacional Soberania, Panama. (**A**) An adult female *E. illicita* and her egg sac on *V. collinsii* near Madden Dam, Panama. Note the patrolling acacia ants on the leaflets and the thorns.

Despite whether or not *E. oblonga* and *E. illicita* are truly host-plant-specific, they are seemingly adapted to inhabiting ant-defended acacias. Patrolling acacia ants regularly encounter the spiders as they rest on the plant surface, often stopping to antennate them before moving on, unperturbed. The spiders typically refrain from reacting to the ants even if the ants walk directly over them. An experiment comparing the reaction of *Pseudomyrmex satanicus* ants on *V. melanoceras* to active versus immobilised *E. oblonga* spiders showed that immobilised spiders did not elicit an aggressive response in the ants. Moving spiders, however, immediately incited ants to become agitated and attack [12]. In response to ant aggression, the spiders either retreated to web strands or, more frequently, ran a short distance and then stopped and crouched against the plant surface, thereby preventing detection by the ants. In contrast, another araneid orb-web species from the surrounding understory used in this experiment, *Argiope argentata*, reacted quite differently. If they were confronted by patrolling ants, instead of running a short distance and then sitting still, they continued to run, further stimulating ant aggression until they were killed or forced off the plant [12]. What do these spiders gain by inhabiting plants patrolled by dangerous ants? In a field experiment in which entire acacia ant colonies were removed from *V. melanoceras* acacias, the abundance of *E. oblonga* spiders decreased significantly over time compared to control acacias. Concomitantly, the abundance of natural enemies of spiders increased on the acacias from which ants were removed, perhaps because they were no longer deterred by patrolling ants. These results suggest *E. oblonga* spiders may be adapted to exploit their hosts' ant–acacia mutualism for enemy-free space [13].

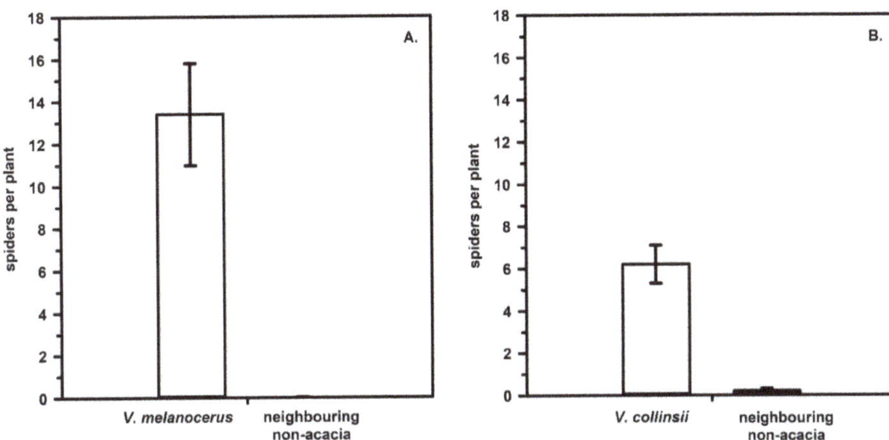

Figure 2. (A). *Eustala oblonga* abundance on *Vachellia melanoceras* acacias and randomly selected neighbouring non-acacias in Parque Nacional Soberania, Panama (ltwo sample t-test: t_{98} = 10.97, $p < 0.0001$). **(B).** *Eustala illicita* abundance on *V. collinsii* acacias and randomly selected neighbouring non-acacias in Parque Natural Metropolitano, Panama (Mann–Whitney U test: U_{18} = 5.6, $p < 0.0001$ from Hesselberg and Triana [11]). Error bars in both panels represent the standard error.

Besides employing behavioural mechanisms to avoid ant aggression, *E. oblonga* and *E. illicita* may also mask their presence on the acacias chemically, either by synthesising or absorbing odours into their cuticles of the *Pseudomyrmex* ant mutualists or their host acacias. Chemical mimicry of host ants has been documented in spider myrmecophiles in a few families that are either predators of ant larvae or kleptoparasites of ant prey (reviewed in Cushing [90]), but such an interaction has not been documented for any araneid spider. Bolas spiders, which are web-less spiders in the family Araneidae, demonstrate that araneid spiders can use chemical mimicry as they emit volatiles that mimic the pheromones of female moths to lure the males close so that they can catch them with their bolas [91].

A preliminary investigation of the chemical mimicry hypothesis provides conflicting evidence for this. In a translocation experiment (K. Marvin and J.D. Styrsky, unpublished data), freshly killed *E. oblonga* and *E. illicita* spiders were moved to either a different individual of their own host acacia species or to non-host acacia species across the Panamanian isthmus, and the time until the spiders were attacked and dragged off the foliage was recorded. The spiders were frozen immediately before being placed on acacias to isolate any effect of chemical camouflage from spider movement that might stimulate ant aggression. Failure-time analyses showed that *E. oblonga* spiders were attacked by patrolling ants significantly more rapidly on non-host acacias (*V. collinsii*) than on their own host acacias (*V. melanoceras*) (Figure 3A). Further, patrolling ants were significantly more likely to lunge at (a confrontational encounter but not an actual attack) *E. oblonga* spiders on non-host acacias than on their host acacias. These results could suggest that *E. oblonga* spiders were 'chemically familiar' to the ants on the spiders' host acacias, but that the ants on the non-host acacias perceived *E. oblonga* spiders as foreign. Contradictory to these results, however, *E. illicita* spiders were no more likely to be lunged at, and were attacked no more frequently on non-host acacias (*V. melanoceras*) than on their host acacias (*V. collinsii*) (Figure 3B). These results are difficult to interpret. At this point, the cuticular chemistry of neither the spiders nor the ants has been analysed to further investigate the chemical mimicry hypothesis.

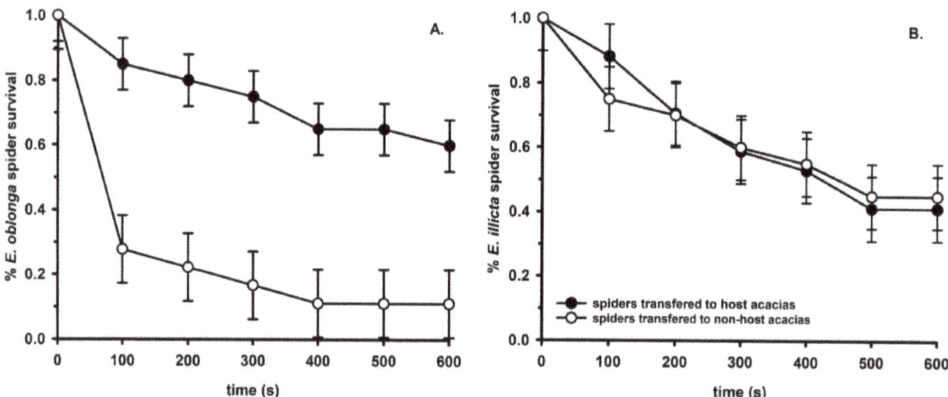

Figure 3. Results of Cox proportional hazards failure-time analyses comparing the percentage of survival over time. (**A**). *E. oblonga* spiders transferred from their home host plant (*V. melanoceras*) to either another host acacia or a non-host acacia (*V. collinsii*) ($X_1^2 = 15.41$, $p < 0.0001$). (**B**). *Eustala illicita* spiders transferred from their home host plant (*V. collinsii*) to either another host acacia or a non-host acacia (*V. melanoceras*) ($X_1^2 = 0.21$, $p = 0.65$). These experiments were conducted in Parque Nacional Soberania and Parque Natural Metropolitano, Panama in 2008. Error bars in both panels represent the standard error.

The cues *Eustala oblonga* and *E. illicita* use to find and discern their respective host acacias from the surrounding understory vegetation are also currently unknown. *Vachellia melanoceras* is sparsely distributed within its range on the Atlantic side of central Panama [89], potentially making it difficult to target. Despite this low density, mature *V. melanoceras* acacias (10–15 m in height) can host hundreds of adult *E. oblonga* spiders (J.D. Styrsky, unpublished data). *Vachellia collinsii* can occur in greater densities in the Pacific side of Panama, but it depends on the particular site [89]. Previous work shows that spiders that are associated with plants use visual, olfactory, and tactile cues to locate specific plant species (reviewed in Vasconcellos-Neto et al. [9]). Given that some spiders are sensitive to plant volatiles, as discussed above, it is possible that *E. oblonga* and *E. illicita* use volatiles produced by the acacias or their acacia ant mutualists to locate host acacias. In a simple choice experiment (D. Clement and J.D. Styrsky, unpublished data), adult *E. oblonga* spiders were offered freshly collected foliage of *V. melanoceras* in one 15.5 cm diameter tube chamber and freshly collected foliage from another understory woody plant haphazardly selected from the immediate vicinity of the acacia in a second chamber. The same experiment was set up to test *E. illicita* spiders using the foliage of its host plant, *V. collinsii*. Individual spiders were placed in a shorter and narrower tube in between the two plant chambers and left for twelve hours. In 13 out of 16 trials, *E. oblonga* spiders were found occupying the acacia foliage (i.e., not just in the chamber with the foliage). Similarly, in 14 out of 16 trials, *E. illicita* were found occupying the acacia foliage. In both experiments, acacia ants had been removed from the acacia foliage before placing the foliage in the chambers, but their cuticular hydrocarbons might still have been detectable.

3.2. The Linyphiid *Laetesia Raveni*

Examples of close spider–plant associations among linyphiids and other web-building families, such as Theridiidae, are rare, and often those described as such do not hold up to closer scrutiny. For example, aside from the previously mentioned *Latrodectus geometricus* (see Nishi [92]; Nishi et al. [73]), *Dipoena banksii* is the only other theridiid reported to have a close plant association. This species is commonly found on *Piper* plants indirectly through its preferred ant prey, which exclusively inhabits *Piper* species [93]. Research is especially limited with regard to linyphiids, and most available accounts of linyphiid–plant interactions are inexplicit, such as the observations by Cresswell [77] and Milne and Waller [78], and a study by Bomfim et al. [75], which recorded an association of two Erigoninae linyphiids with the thorny rosette-structured plant *Eryngium horridum*. Thorny plants can provide important microhabitats for some web-building spider species, as demonstrated by Ruch et al. [72] and Rose et al. [54]. The thorns create a complex architecture that provides sufficient web attachment sites whilst simultaneously protecting the spiders from disturbances or threats [54,58,72,74]. Interestingly, a new species of Linyphiidae, *Laetesia raveni* (Figure 4), recently described by Hormiga and Scharff [10], has been observed exclusively on two thorny plant species, *Calamus muelleri* and *Solanum inaequilaterum* in Queensland, Australia. The unique case of *L. raveni* is currently the only recorded instance of a linyphiid exhibiting host plant specificity. This linyphiid constructs a dome-shaped web on its chosen host plant, and according to Hormiga and Scharff [10], the spiders were typically situated upside-down on the web, directly under a leaf that was positioned at the centre of the web. Often, *L. raveni* was observed flattening its body against the leaf when provoked. The authors suggest that this behaviour, combined with the spider's unique green colouration, is a form of crypsis (Figure 4). *Laetesia raveni* was more common on *C. muelleri*, a climbing palm with stems densely covered with thorns, and with the leaflets, stalks, and midribs of the fronds also bearing small spines. Similarly, *S. inaequilaterum* has thorns or spines covering the stem and leaves; however, the thorns are much denser along the stem. There have been two recorded instances where *L. raveni* was found on other undocumented plant species, but these plants were seemingly in physical contact with either *C. muelleri* or *S. inaequilaterum* [10]. The ecology of *L. raveni* and its unusual association with *C. muelleri* and *S. inaequilaterum* requires considerably more research, especially to: (1) Determine if this is a host-specific association (unpublished observations from the rainforest reserve in Lismore, New South Wales suggest that it can be found on other plant species (N. Fisher, personal communication October 2022, Figure 4)); (2) Further understand whether *L. raveni* inhabits the host plants for protection from natural enemies, as seen in *Stegodyphus* species (see Ruch et al. [72]; Rose et al. [54]); and (3) Determine if the green colouration and body-flattening behaviour are forms of passive predator defence (i.e., crypsis).

Figure 4. Photographs of *Laetesia raveni* (Linyphiidae) showing its close association with vegetation. (**A**). Frontal view. Photo taken by Samuel Frankel from iNaturalist. (**B**). Dorsal view demonstrating flattening against the leaf of *Mallotus philippinensis* in a possible camouflage attempt. Photo courtesy of Nick Fisher from Flickr (dustaway).

4. How Might Spiders Identify and Locate Their Host Plant?

4.1. Website Choices and Web Building

Web-building behaviour is relatively well studied, especially in tangle-(Theridiidae) and orb-web spiders (Araneidae and Tetragnathidae), where it follows a generally rigid pattern of stereotypic behaviours, although with some flexibility [94,95]. On the other hand, however, we still know very little about the process of habitat exploration and site selection that precedes web construction [96,97]. Orb-web spiders engage in extensive site exploration [98], and generally match the shape of their webs to both the available space [99] and the available silk supplies [100]. Most web-building spiders are not picky when it comes to attaching their webs to their surroundings and will choose any suitable structure—usually a rigid or semi-rigid structure in order to avoid web damage from wind movements [5,101,102]—although some spiders also attach their webs to moving structures, such as leaves and grass, without it negatively affecting their web-building efficiency or resultant webs [103]. Linyphiid spiders in grassland, for example, do not show any preference for specific plant species, but consistently select tall and stable vegetation to attach their webs to [5]. Similarly, the desert-inhabiting social eresid spiders, *Stegodyphus dumicola*, construct their colonial webs on taller rigid plants with thorns [54]. Individual spiders seem to select optimal host plants based on the structural properties of the plant, including their fractal dimension [104], while some web-building spiders select web-attachment sites on substrate depending on its hydrophobicity [105].

We know virtually nothing about how the few web-building spiders with specific host plants choose them, but interestingly even these associations can be flexible. For example, as discussed above, the acacia orb-web spider *Eustala illicita* is almost exclusively found on the acacia *Vachellia collinsii* with only four juveniles out of a total of 117 observed spiders found in neighbouring vegetation. It nonetheless readily builds webs in sterile plastic frames in the laboratory [11]. All age classes, from early juveniles to adult females, build webs in captivity at high web-building frequencies with the webs being, at least superficially, very similar to the ones built in the wild [11,106]. Spiders in the lab also show a high degree of flexibility in adapting their web shape to differently shaped plastic frames [99]. No learning seemed to be involved as the second and third webs constructed in the frames are no different than the first web [107], which suggests that this flexibility is regularly needed in their natural habitats. This fits well with the observation that acacia spiders can be found in high densities on their acacia host plants [13], which presumably gives rise to competition over suitable web-building spaces forcing some spiders to build at less optimal sites within the tree, where adaptations to the standard orb web shape and structure are required. In the case of the web-building spiders that are not closely associated with specific species of plants, and possibly also for those few that are, suitably structured vegetation for building webs is probably found by random searching and mechanical contact stimulation, as web-building spiders typically have very poor vision [108].

Many spiders engage in random dispersal through ballooning either as adults (if small spiders) or in the early juvenile stages. This involves releasing silk threads into the wind, where a combination of electric and aerodynamic forces lift the spider into the air and potentially disperse it over long distances [109,110]. Ballooning is common in web-building spiders (e.g., [111–114]). Ballooning propensity is highest in spiders living in open ecosystems, although one study found that some spiders from temperate woodlands can have high ballooning propensity similar to those of grassland [115]. To our knowledge, however, no data are available on ballooning propensity in web-building spiders in tropical rainforests, so it remains currently unknown if spiders associating with specific tropical forest trees, such as the acacia orb-web spiders in the genus *Eustala*, discussed in this paper, use short distance ballooning as a host plant location strategy.

Spiders are known to use short-distance random or systematic search strategies for locating lost egg sacs and caught prey with examples of the former from cursorial spiders [116] and the latter from web-building spiders [117,118]. On the other hand, while some spiders are known to be able to find their burrows over long distances, likely using compass and path integration [119,120], no information, at least not to our knowledge, is available on the extent to which spiders rely on random-search patterns to find suitable web-building sites over longer distances. However, the hypothesis that many spiders engage in random searching and potential trial-and-error web-building behaviour on chosen sites is strengthened by the observations that some orb-web spiders, despite extensive site exploration prior to web-building [97], build a smaller explorative web when building at a new location [98], and readily move their webs when encountering low prey capture, or when suffering web damage [70,121,122].

4.2. The Use of Chemical Cues and Communication in Spiders

The alternative to the random or systematic search strategies for finding suitable plant hosts discussed above is a more targeted strategy using chemical cues. Spiders are known to use chemical cues in sexual communication, especially in relation to males locating females through silk-borne [123] or cuticular cues [124]. We refer readers to the recent excellent review by Fischer [19] on chemical communication in spiders focusing on a methodological overview on how to study their pheromones. Spiders can also detect predators such as ants through semiochemicals [125], and they are sensitive to the chemical cues of potential predators [126,127]. The wolf spider *Pardos milvina*, for example, alters where it forages when it chemically detects one of its predators, the larger wolf spider *Trigrosa helluo* [126].

Many spiders associated with ants use chemical cues from the ants to prey upon them. The mimicry of cuticular hydrocarbons (CHCs) is recognized as one of the most common mechanisms that myrmecophiles and termitophiles use to deceive their host [128] but evidence from spiders is scarce. The jumping spider *Cosmophasis bitaeniata* uses the CHC mimicry of its ant host *Oecophylla smaragdina* to prey on larvae [129]. Interestingly, the spider does not acquire the chemical mimicry by physical contact with the adult ants, but it acquires it from eating the larvae, and the variation in CHCs profiles across spiders is colony-specific [130]. Another foraging strategy is chemical eavesdropping, as in the myrmecophagous jumping spider *Habrocestum pulex*, which initiates predatory behaviours when presented with airborne and soilborne chemical cues from the ants [21]. Chemical eavesdropping can show phenotypic plasticity, as seen in the jumping spider *Cyrba algerina*, which varies its responsiveness towards spider prey odours depending on whether the prey species cohabits with the spider or not [131]. Eavesdropping can also be used by ant-mimicking spiders to find their mimetic model ant species (Batesian mimicry) without preying on the ants [132]. In other instances, spiders seem to choose a habitat that increases their chances of foraging success. For example, the western black widow *Latrodectus hesperus* prefers to build its webs in microhabitats where it detects the residual chemical cues of house crickets [133].

As discussed in the 'spider–plant associations' section above, spider–plant interactions are now widely described and plants with rosette-shaped clusters of leaves or tri-chomes are the most common plant architectures to have associations with spiders [9]. The evidence that some spiders select plants with similar architectural features by using visual cues is strong (reviewed in Vasconcellos-Neto et al. [9]), but few studies have explored whether chemical signals are involved in host plant recognition. However, the examples are known to include the nursery web spider [134], crab spiders [22,135], and jumping spiders [136]. For instance, pitfall traps baited with eugenol—which is a flower component fragrance—caught more individuals of two *Thomisus* species (Thomisidae), as compared to controls [22]. Similarly, *Thomisus spectabilis* chose the same flower more often than a honeybee, when there was a flower odour signal present [135]. Interestingly, chemical recognition of the host plant is species-specific in some cases, with some plant chemicals inducing responses in some spiders but not in others [134,137]. Thus, it remains a possibility that the *Eustala* spiders use plant volatiles to locate their hosts.

5. Conclusions and Future Directions

We found, despite extensive literature searches, only two examples of web-building spiders showing host specificity, and even in these two examples, some individuals were observed on non-host plants. Our study, therefore, suggests that web-building spiders in general are less likely to form one-to-one associations with specific species of plants than cursorial spiders (see also Vasconcellos-Neto et al. [9]). One reason for this could be due to a lack of research focusing on looking for these relationships in web-building spiders, but it is likely that they are in fact rare since web-building spiders create their own modified microhabitat with the web acting like an extended phenotype [138]. If the surrounding vegetation is reduced to just providing support or shelter for the web [97,139], it stands to reason that web-building spiders have a far less intimate relationship with the vegetation than the cursorial spiders that spend most of their life in direct contact with one or a few species of plants.

More studies on host plant and web-site preferences are needed in web-building spiders, especially of smaller spiders, such as many linyphiids and some theridiids, which construct small webs fully within a single plant. Undoubtedly, more examples of host specificity in web-building spiders await discovery, especially from tropical regions where the diversity of these spiders is highest. We also need more detailed studies on the known interactions as many unanswered questions remain, including whether spiders that use several host plant species [57,61] rely on the physical attributes of the plants or if instead these plants share similar chemical profiles or particular molecules that facilitate recognition. Spiders that specialise in certain plant families as host plants [61] are good candidates to address these questions. Furthermore, ontogenetic variation should be integrated into the study of spider chemical ecology. For instance, plant specialisation in the Japanese foliage spider, *Cheiracanthium japonicum*, seems to develop with age, with juveniles and adults using different plant species in some cases [140].

In the present study, we found one very interesting and well-evidenced example coming from the acacia orb-spiders in the araneid genus *Eustala*, which seem to use the acacia and their ant protectors for enemy-free space [13] without causing any significant harm to either the plants or the ants [89]. The particularly interesting aspect of this example is that we currently know of two species of *Eustala* that are associated with two different species of *Vachellia* with their own specific species of *Pseudomyrmex* ants [11,12]. While a few individuals have been found in nearby vegetation, particularly in dead vegetation as is common in other species of *Eustala* [86,88], this indicates a high degree of host specificity, probably aided by a combination of behavioural and chemical mimicry to avoid attacks from the resident ants [12]. These findings and the preliminary data we discuss in Section 3.1 above suggest that both *Eustala* species may use chemical cues to discern host acacias. However, to confirm this hypothesis, larger and more detailed studies on both short-distance (centimetre scale) chemical attraction in laboratory behavioural assays with Y- or T-mazes, and longer distance (meter scale) navigation in the laboratory and in the field are needed to determine the potential role of plant volatiles and/or ant pheromones for host location identification. Similarly, we need a combination of behavioural and cuticular chemistry studies (such as comparing surface chemistry profiles in spiders, ants, and acacias with GG-MS) to determine the degree to which *E. illicita* and *E. oblonga* rely exclusively on measured behavioural responses [12] to avoid getting attacked by the aggressive *Pseudomyrmex* ants.

The scattered distribution of acacias within the rainforest could also suggest that they can be viewed as habitat islands from the perspective of the spiders and insects that utilise the ant–acacia system [141]. Thus, studies on the mechanisms behind targeted navigation and host-finding mechanisms could be combined with studies on gene flow between spider populations on individual trees or groups of trees in different parts of the same forest. DNA sequence differences, usually from mitochondrial genes, can be used to determine pairwise F_{ST} differences among samples collected at different geographic scales [142]. This method has been successfully used with orb-web spiders several times, including Lee and co-workers' [143] study revealing a high level of gene flow between *Nephila pilipes* populations across a mountain range in Taiwan. The surprisingly high interconnectedness between these spatially isolated populations is almost certainly caused by long distance dispersal via ballooning, which many spiders engage in [144]. It is currently not known if, or to what degree, *Eustala* orb spiders engage in ballooning, but studies on the propensity of ballooning, which can easily be quantified in the laboratory [112], could be fruitfully combined with studies on gene structure to further cast light on the intimate relationships between these spiders and their host plants.

The wide range of questions that can be asked and answered in the spider–ant–acacia system indicate that this system makes for an ideal model system for evolutionary and ecological studies, especially as comparative studies can be conducted on different closely related species and because strategies can be contrasted with other arthropods that utilise the swollen thorn acacias for enemy-free space, or engage in parasitic interactions with either the ants or the acacias [145,146].

Author Contributions: Conceptualization, T.H. and D.G.; writing original draft, all authors; writing—review and editing, all authors; visualization T.H., K.M.B. and J.D.S.; funding acquisition, T.H., J.D.S. and D.G. All authors have read and agreed to the published version of the manuscript.

Funding: The authors gratefully acknowledge funding from the Panamanian government (SENACYT grant FID22-034), Sistema Nacional de Investigación, a Royal Society International Exchanges grant (IES\R3\213007) and grants from the Thomas F. Jeffress and Kate Miller Jeffress Memorial Trust, the Virginia Foundation for Independent Colleges and the University of Lynchburg.

Acknowledgments: Previously unpublished data presented in this manuscript were collected under Autoridad Nacional del Ambiente permits SE/A-56-08 and SE/A-43-09. We would also like to thank three anonymous reviewers for their detailed comments on content and structure, which significantly improved the paper.

Conflicts of Interest: The authors declare no conflict of interest.

References

1. Janzen, D.H. Coevolution of mutualism between ant and acacias in Central America. *Evolution* **1966**, *20*, 249–274. [CrossRef] [PubMed]
2. Weiblen, G.D. How to be a fig wasp. *Annu. Rev. Entomol.* **2002**, *47*, 299–330. [CrossRef] [PubMed]
3. Pauw, A. Floral syndromes accurately predict pollination by a specialized oil-collecting bee (*Rediviva peringueyi*, Melittidae) in a guild of South African orchids (Coryciinae). *Reprod. Biol.* **2006**, *93*, 917–926. [CrossRef] [PubMed]
4. Vacante, V. *The Handbook of Mites of Economic Plants: Identification, Bio-Ecology and Control*, 1st ed.; CAB International: Oxford, UK, 2016. [CrossRef]
5. Gibson, C.W.; Hambler, C.; Brown, V.K. Changes in spider (Araneae) assemblages in relation to succession and grazing management. *J. Appl. Ecol.* **1992**, *29*, 132–142. [CrossRef]
6. Patt, J.M.; Pfannenstiel, R.S. Odor-based recognition of nectar in cursorial spiders. *Entomol. Exp. Appl.* **2008**, *127*, 801–809. [CrossRef]
7. Meehan, C.J.; Olson, E.J.; Reudink, M.W.; Kyser, T.K.; Curry, R.L. Herbivory in a spider through exploitation of an ant-plant mutualism. *Curr. Biol.* **2009**, *19*, 892–893. [CrossRef]
8. Eggs, B.; Sanders, D. Herbivory in spiders: The importance of pollen for orb-weavers. *PLoS ONE* **2013**, *8*, e82637. [CrossRef]
9. Vasconcellos-Neto, J.; Messas, Y.F.; Souza, H.D.S.; Villanueva-Bonilla, G.A.; Romero, G.Q. Spider-plant interactions: An ecological approach. In *Behaviour and Ecology of Spiders: Contributions from the Neotropical Region*, 1st ed.; Viera, C., Gonzaga, M.O., Eds.; Springer: Cham, Switzerland, 2017; pp. 165–214. [CrossRef]
10. Hormiga, G.; Scharff, N. The strange case of *Laetesia raveni* n. sp., a green linyphiid spider from Eastern Australia with a preference for thorny plants (Araneae, Linyphiidae). *Zootaxa* **2014**, *3811*, 83–94. [CrossRef]
11. Hesselberg, Y.; Triana, E. The web of the acacia orb-spider *Eustala illicita* (Araneae: Araneidae) with notes on its natural history. *J. Arachnol.* **2010**, *38*, 2126. [CrossRef]
12. Garcia, L.C.; Styrsky, J.D. An orb-weaver spider eludes plant-defending acacia ants by hiding in plain sight. *Ecol. Entomol.* **2013**, *28*, 190–199. [CrossRef]
13. Styrsky, J.D. An orb-weaver spider exploits an ant–acacia mutualism for enemy-free space. *Ecol. Evol.* **2014**, *4*, 276–283. [CrossRef]
14. Bruce, T.J.A.; Wadhams, L.J.; Woodcock, C.M. Insect host location: A volatile situation. *Trends Plant Sci.* **2005**, *10*, 269–274. [CrossRef]

15. Wang, H.M.; Bai, P.H.; Zhang, J.; Zhang, X.M.; Hui, Q.; Zheng, H.X.; Zhang, X.H. Attraction of bruchid beetles *Callosobruchus chinensis* (L.) (Coleoptera: Bruchidae) to host plant volatiles. *J. Integr. Agric.* **2020**, *19*, 3035–3044. [CrossRef]

16. Blatrix, R.; Mayer, V. Communication in ant-plant symbioses. In *Plant Communication from an Ecological Perspective: Signaling and Communication in Plants*, 1st ed.; Baluška, F., Ninkovic, V., Eds.; Springer: Heidelberg/Berlin, Germany, 2010.

17. Dáttilo, W.F.C.; Izzo, T.J.; Inouye, B.D.; Vasconcelos, H.L.; Bruna, E.M. Recognition of host plant volatiles by *Pheidole minutula* Mayr (myrmicinae), an Amazonian ant-plant specialist. *Biotropica* **2009**, *41*, 642–646. [CrossRef]

18. Inui, Y.; Itioka, T.; Murase, K.; Yamaoka, R.; Itino, T. Chemical recognition of partner plant species by foundress ant queens in *Macaranga-Crematogaster* myrmecophytism. *J. Chem. Ecol.* **2001**, *27*, 2029–2040. [CrossRef] [PubMed]

19. Fischer, A. Chemical communication in spiders—A methodological review. *J. Arachnol.* **2019**, *47*, 1–27. [CrossRef]

20. Scott, C.; Gerak, C.; McCann, S.; Gries, G. The role of silk in courtship and chemical communication of the false widow spider, *Steatoda grossa* (Araneae: Theridiidae). *J. Ethol.* **2018**, *36*, 191–197. [CrossRef]

21. Clark, R.J.; Jackson, R.R.; Cutler, B. Chemical cues from ants influence predatory behavior in *Harbocestum pulex* (Araneae, Salticidae). *J. Arachnol.* **2000**, *28*, 309318. [CrossRef]

22. Krell, F.K.; Krämer, F. Chemical attraction of crab spiders (Araneae, Thomisidae) to a flower fragrance component. *J. Arachnol.* **1998**, *26*, 117–119.

23. Rao, D. Habitat selection and dispersal. In *Behaviour and Ecology of Spiders: Contributions from the Neotropical Region*, 1st ed.; Viera, C., Gonzaga, M.O., Eds.; Springer: Cham, Switzerland, 2017; pp. 85–108. [CrossRef]

24. Young, O.P.; Lockley, T.C. Spiders of Spanish moss in the Delta of Mississippi. *J. Arachnol.* **1989**, *17*, 143–148.

25. Romero, G.Q.; Vasconcellos-Neto, J. Spatial distribution patterns of jumping spiders associated with terrestrial bromeliads. *Biotropica* **2004**, *36*, 596–601. [CrossRef]

26. Romero, G.Q. Geographic range, habitats and host plants of bromeliad-living jumping spider (Salticidae). *Biotropica* **2006**, *38*, 522–530. [CrossRef]

27. Romero, G.Q.; Vasconcellos-Neto, J. Foraging by the flower-dwelling spider, *Misumenops argenteus* (Thomisidae), at high prey density sites. *J. Nat. Hist.* **2004**, *38*, 1287–1296. [CrossRef]

28. Romero, G.Q.; Vasconcellos-Neto, J. Spatial distribution and microhabitat preference of *Psecas chapoda* (Peckham & Peckham) (Araneae, Salticidae). *J. Arachnol.* **2005**, *33*, 124–134.

29. Romero, G.Q.; Vasconcellos-Neto, J. Population dynamics, age structure and sex ratio of the bromeliad-dwelling jumping spider, Psecas chapoda (Salticidae). *J. Nat. Hist.* **2005**, *39*, 153–163. [CrossRef]

30. Gonçalves, A.Z.; Mercier, H.; Mazzafera, P.; Romero, G.Q. Spider-fed bromeliads: Seasonal and interspecific variation in plant performance. *Ann. Bot.* **2011**, *107*, 1047–1055. [CrossRef] [PubMed]

31. Romero, G.Q.; Mazzafera, P.; Vasconcellos-Neto, J.; Trivelin, P.C.O. Bromeliad-living spiders improve host plant nutrition and growth. *Ecology* **2006**, *87*, 803–808. [CrossRef]

32. Romero, G.Q.; Vasconcellos-Neto, J.; Trivelin, P.C.O. Spatial variation in the strength of mutualism between a jumping spider and a terrestrial bromeliad: Evidence from the stable isotope 15N. *Acta Oecologica* **2008**, *33*, 380–386. [CrossRef]

33. Thorogood, C.J.; Bauer, U.; Hiscock, S.J. Convergent and divergent evolution in carnivorous pitcher plant traps. *New Phytol.* **2018**, *217*, 1035–1041. [CrossRef]

34. Lim, R.J.Y.; Lam, W.N.; Tan, H.T.W. Novel pitcher plant–spider mutualism is dependent upon environmental resource abundance. *Oecologia* **2018**, *188*, 791–800. [CrossRef]

35. Choo, J.P.S.; Koh, T.L.; Ng, P.K.L. Pitcher fluid macrofauna: Nematodes and arthropods. In *A Guide to the Carnivorous Plants of Singapore*, 1st ed.; Tan, H.T.W., Ed.; Singapore Science Centre: Singapore, 1997.

36. Rembold, K.; Fischer, E.; Striffler, B.F.; Barthlott, W. Crab spider association with the Malagasy pitcher plant *Nepenthes madagascariensis*. *Afr. J. Ecol.* **2012**, *51*, 188–191. [CrossRef]

37. Lam, W.N.; Tan, H.T.W. The crab spider–pitcher plant relationship is a nutritional mutualism that is dependent on prey-resource quality. *J. Anim. Ecol.* **2019**, *88*, 102–113. [CrossRef] [PubMed]

38. Anderson, B.; Midgely, J.J. It takes two to tango but three is a tangle: Mutualists and cheaters on the carnivorous plant *Roridula*. *Oecologia* **2002**, *132*, 369–373. [CrossRef] [PubMed]

39. Morais-Filho, J.C.; Romero, G.Q. Microhabitat use by *Peucetia flava* (Oxyopidae) on the glandular plant *Rhyncanthera dichotoma* (Melastomataceae). *J. Arachnol.* **2008**, *36*, 374–378. [CrossRef]

40. Morais-Filho, J.C.; Romero, G.Q. Plant glandular trichomes mediate protective mutualism in a spider–plant system. *Ecol. Entomol.* **2010**, *35*, 485–494. [CrossRef]

41. Gavini, S.S.; Quintero, C.; Tadey, M. Ecological role of a flower-dwelling predator in a tri-trophic interaction in northwestern Patagonia. *Acta Oecologica* **2019**, *95*, 100–107. [CrossRef]

42. Vasconcellos-Neto, J.; Romero, G.Q.; Santos, A.J.; Dippenaar-Schoeman, A.S. Associations of spiders of the genus *Peucetia* (Oxyopidae) with plants bearing glandular hairs. *Biotropica* **2007**, *39*, 221–226. [CrossRef]

43. Kessler, A.; Baldwin, I.T. Plant responses to insect herbivory: The emerging molecular analysis. *Annu. Rev. Plant Biol.* **2002**, *53*, 299–328. [CrossRef]

44. Wagner, G.J.; Wang, E.; Shepherd, R.W. New approaches for studying and exploiting an old protuberance, the plant trichome. *Ann. Bot.* **2004**, *93*, 3–11. [CrossRef]

45. Sousa-Lopes, B.; Alves-da-Silva, N.; Alves-Martins, F.; Del-Claro, K. Antiherbivore protection and plant selection by the lynx spider *Peucetia flava* (Araneae: Oxyopidae) in the Brazilian Cerrado. *J. Zool.* **2019**, *308*, 121–127. [CrossRef]
46. Romero, G.Q.; Souza, J.C.; Vasconcellos-Neto, J. Anti-herbivore protection by mutualistic spiders and the role of plant glandular trichomes. *Ecology* **2008**, *89*, 3105–3115. [CrossRef] [PubMed]
47. Morais-Filho, J.C.; Romero, G.Q. Natural history of *Peucetia flava* (Araneae, Oxyopidae): Seasonal density fluctuation, phenology and sex ratio on the glandular plant *Rhyncanthera dichotoma* (Melastomataceae). *J. Nat. Hist.* **2009**, *43*, 701–711. [CrossRef]
48. Gómez-Acevedo, S.L. New reports of spiders in three Mexican ant-acacias. *Open J. Ecol.* **2021**, *11*, 32–40. [CrossRef]
49. Heiling, A.M.; Chittka, L.; Cheng, K.; Herberstein, M.E. Colouration in crab spiders: Substrate choice and prey attraction. *J. Exp. Biol.* **2005**, *208*, 1785–1792. [CrossRef]
50. Llandres, A.L.; Gawryszewski, F.M.; Heiling, A.M.; Herberstein, M.E. The effect of colour variation in predators on the behaviour of pollinators: Australian crab spiders and native bees. *Ecol. Entomol.* **2011**, *36*, 72–81. [CrossRef]
51. Théry, M.; Debut, M.; Gomez, D.; Casas, J. Specific color sensitivities of prey and predator explain camouflage in different visual systems. *Behav. Ecol.* **2005**, *16*, 25–29. [CrossRef]
52. Su, Q.; Qi, L.; Zhang, W.; Yun, Y.; Zhao, Y.; Peng, Y. Biodiversity survey of flower-visiting spiders based on literature review and field study. *Environ. Entomol.* **2020**, *49*, 673–682. [CrossRef]
53. Brechbühl, R.; Casas, J.; Bacher, S. Ineffective crypsis in a crab spider: A prey community perspective. *Proc. Roy. Soc. B* **2009**, *277*, 739–746. [CrossRef]
54. Rose, C.; Schramm, A.; Irish, J.; Bilde, T.; Bird, T.L. Host plant availability and nest-site selection of the social spider *Stegodyphus dumicola* Pocock, 1898 (Eresidae). *Insects* **2022**, *13*, 30. [CrossRef]
55. Romero, G.Q.; Vasconcellos-Neto, J. The effects of plant structure on the spatial and microspatial distribution of a bromeliad-living jumping spider (Salticidae). *J. Anim. Ecol.* **2005**, *74*, 12–21. [CrossRef]
56. Omena, P.M.; Romero, G.Q. Fine-scale microhabitat selection in a bromeliad-dwelling jumping spider (Salticidae). *Biol. J. Linn. Soc.* **2008**, *94*, 653–662. [CrossRef]
57. Jacobucci, G.B.; Medeiros, L.; Vasconcellos-Neto, J.; Romero, G.Q. Habitat selection and potential antiherbivore effects of *Peucetia flava* (Oxyopidae) on *Solanum thomasiifolium* (Solanaceae). *J. Arachnol.* **2009**, *37*, 365–367. [CrossRef]
58. Uetz, G.W. Habitat Structure and Spider Foraging. In *Habitat Structure*, 1st ed.; Bell, S.S., McCoy, E.D., Mushinsky, H.R., Eds.; Springer: Dordrecht, Switzerland, 1991; pp. 325–348. [CrossRef]
59. Wise, D.H. *Spiders in Ecological Webs*, 1st ed.; Cambridge University Press: Cambridge, UK, 1993. [CrossRef]
60. Souza, A.L.T.D.; Martins, R.P. Distribution of plant-dwelling spiders: Inflorescences versus vegetative branches. *Austral Ecol.* **2004**, *29*, 342–349. [CrossRef]
61. Villanueva-Bonilla, G.A.; Salomão, A.T.; Vasconcellos-Neto, J. Trunk structural traits explain habitat use of a tree-dwelling spider (Selenopidae) in a tropical forest. *Acta Oecologica* **2017**, *85*, 108–115. [CrossRef]
62. Arango, A.M.; Rico-Gray, V.; Parra-Tabla, V. Population structure, seasonality, and habitat use by the green lynx spider *Peucetia viridans* (Oxyopidae) inhabiting *Cnidoscolus aconitifolius* (Euphorbiaceae). *J. Arachnol.* **2000**, *28*, 185–194. [CrossRef]
63. Corcuera, P.; Valverde, P.L.; Jiménez-Salinas, E.; Vite, F.; López-Ortega, G.; Pérez-Hernández, M.A. Distribution of *Peucetia viridans* (Araneae: Oxyopidae) on Croton ciliatoglandulifer. *Environ. Entomol.* **2010**, *39*, 320–327. [CrossRef] [PubMed]
64. Cross, F.R. Attentional Processes in Mosquito-Eating Jumping Spiders: Search Images and Cross-Modality Priming. Ph.D. Dissertation, University of Canterbury, Canterbury, New Zealand, 2009.
65. Agnarsson, I.; Rayor, L.S. A molecular phylogeny of the Australian huntsman spiders (Sparassidae, Deleninae): Implications for taxonomy and social behaviour. *Mol. Phyl. Evol.* **2013**, *69*, 895–905. [CrossRef]
66. Quintero, C.; Corley, J.C.; Aizen, M.A. Weak trophic links between a crab-spider and the effective pollinators of a rewardless orchid. *Acta Oecologia* **2015**, *62*, 32–39. [CrossRef]
67. Karl, I.; Bauer, U. Inside the trap. Biology and behavior of the pitcher-dwelling crab spider, *Misumenops nepenthicola*. *Plants People Planet* **2020**, *2*, 290–293. [CrossRef]
68. Anderson, B. Inferring evolutionary patterns from the biogeographical distributions of mutualists and exploiters. *Biol. J. Linn. Soc.* **2006**, *89*, 541–549. [CrossRef]
69. Tanaka, K. Energetic cost of web construction and its effect on web relocation in the web-building spider *Agelena limbata*. *Oecologia* **1989**, *81*, 459–464. [CrossRef] [PubMed]
70. Chmiel, K.; Herberstein, M.E.; Elgar, M.A. Web damage and feeding experience influence web site tenacity in the orb-web spider *Argiope keyserlingi* Karsch. *Anim. Behav.* **2000**, *60*, 821–826. [CrossRef] [PubMed]
71. Whitney, K.D. Experimental evidence that both parties benefit in a facultative plant-spider mutualism. *Ecology* **2004**, *85*, 1642–1650. [CrossRef]
72. Ruch, J.; Heinrich, L.; Bilde, T.; Schneider, J.M. Site selection and foraging in the eresid spider *Stegodyphus tentoriicola*. *J. Insect Behav.* **2012**, *25*, 1–11. [CrossRef]
73. Nishi, A.H.; Vasconcellos-Neto, J.; Romero, G.Q. The role of multiple partners in a digestive mutualism with a protocarnivorous plant. *Ann. Bot.* **2013**, *111*, 143–150. [CrossRef] [PubMed]
74. Gómez, J.E.; Lohmiller, J.; Joern, A. Importance of vegetation structure to the assembly of an aerial web-building spider community in North American open grassland. *J. Arachnol.* **2016**, *44*, 28–35. [CrossRef]

75. Bomfim, L.D.S.; Bitencourt, J.A.G.; Rodrigues, E.N.L.; Podgaiski, L.R. The role of a rosette-shaped plant (*Eryngium horridum*, Apiaceae) on grassland spiders along a grazing intensity gradient. *Insect Conserv. Divers* **2021**, *14*, 492–503. [CrossRef]

76. Jolivet, P.; Vasconcellos-Neto, J. Convergence chez les plantes carnivores. *La Rech.* **1993**, *24*, 456–458.

77. Cresswell, J.E. Capture rates and composition of insect prey of the pitcher plant Sarracenia purpurea. *Am. Midl. Nat.* **1991**, *125*, 1–9. [CrossRef]

78. Milne, M.A.; Waller, D.A. Does pitcher plant morphology affect spider residency? *Northeast. Nat.* **2013**, *20*, 419–429. [CrossRef]

79. Fage, L. Au sujet de deux araignées nouvelles trouvées dans les urnes de *Nepenthes*. *Treubia* **1930**, *12*, 23–28.

80. Lubin, Y.; Hennicke, J.; Schneider, J. Settling decisions of dispersing *Stegodyphus lineatus* (Eresidae) young. *Isr. J. Ecol. Evol.* **1998**, *44*, 217–225.

81. Cuff, J.P.; Evans, S.A.; Porteous, I.A.; Quiñonez, J.; Evans, D.M. Candy-striped spider leaf and habitat preferences for egg deposition. *Agric. For. Entomol.* **2022**, *24*, 422–431. [CrossRef]

82. Chickering, A.M. The genus *Eustala* (Araneae, Argiopidae) in Central America. *Bull. Mus. Comp. Zool.* **1955**, *112*, 391–518.

83. Poeta, M.R.M. The orb-weaving spider genus *Eustala* Simon, 1895 (Araneae, Araneidae): Eight new species, redescriptions, and new records. *Zootaxa* **2014**, *3872*, 440–446. [CrossRef] [PubMed]

84. World Spider Catalog: NMBE. Available online: http://wsc.nmbe.ch (accessed on 29 October 2022).

85. Comstock, J.H. *The Spider Book*, 1st ed.; Doubleday, Page and Company: New York, NY, USA, 1912.

86. Levi, H.W. The American orb-weaver genera *Cyclosa*, *Metazygia* and *Eustala* north of Mexico (Araneae, Araneidae). *Bull. Mus. Comp. Zool.* **1997**, *148*, 61–127.

87. Messas, Y.F.; Souza, H.S.; Gonzaga, M.O.; Vasconcellos-Neto, J. Spatial distribution and substrate selection by the orb-weaver spider *Eustala perfida* Mello-Leitão, 1947 (Araneae: Araneidae). *J. Nat. Hist.* **2014**, *48*, 2645–2660. [CrossRef]

88. Souza, H.S.; Messas, Y.F.; Gonzaga, M.O.; Vasconcellos-Neto, J. Substrate selection and spatial segregation by two congeneric species of *Eustala* (Araneae: Araneidae) in southeastern Brazil. *J. Arachnol.* **2015**, *43*, 59–66. [CrossRef]

89. Ledin, A.E.; Styrsky, J.D.; Styrsky, J.N. Friend or foe? orb-weaver spiders inhabiting ant-acacias capture both herbivorous insects and acacia ant alates. *J. Insect Sci.* **2020**, *20*, 1–8. [CrossRef]

90. Cushing, P.E. Spider-ant associations: An updated review of myrmecomorphy, myrmecophily, and myrmecophagy in spiders. *Psyche* **2012**, *2012*, 151989. [CrossRef]

91. Eberhard, W.G. Aggressive chemical mimicry by a bolas spider. *Science* **1977**, *198*, 1173–1175. [CrossRef] [PubMed]

92. Nishi, A.H. Mutualismo Digestivo Entre Aranhas, Cupins e a Planta Protocarnívora *Paepalanthus bromelioides* (Eriocaulaceae). Master's Thesis, São Paulo State University (Universidade Estadual Paulista), São Paulo, Brazil, 2011.

93. Gastreich, K.R. Trait-mediated indirect effects of a theridiid spider on an ant-plant mutualism. *Ecology* **1999**, *80*, 1066–1070. [CrossRef]

94. Zschokke, S.; Vollrath, F. Unfreezing the behavior of two orb spiders. *Physiol. Behav.* **1995**, *58*, 1167–1173. [CrossRef] [PubMed]

95. Benjamin, S.P.; Zschokke, S. Webs of theridiid spiders: Construction, structure and evolution. *Biol. J. Linn. Soc.* **2003**, *78*, 293–305. [CrossRef]

96. Vollrath, F. Analysis and interpretation of orb spider exploration and web-building behavior. *Adv. Study. Behav.* **1992**, *21*, 147–199. [CrossRef]

97. Hesselberg, T. Exploration behaviour and behavioural flexibility in orb-web spiders: A review. *Curr. Zool.* **2015**, *61*, 313–327. [CrossRef]

98. Zschokke, S. Early stages of orb web construction in *Araneus diadematus* Clerck. *Rev. Suisse. Zool.* **1996**, *2*, 709–720.

99. Hesselberg, T. Web-building flexibility differs in two spatially constrained orb spiders. *J. Insect Behav.* **2013**, *26*, 283–303. [CrossRef]

100. Eberhard, W.G. Behavioral flexibility in orb web construction: Effects of supplies in different silk glands and spider size and weight. *J. Arachnol.* **1988**, *16*, 295–302.

101. Craig, C.L. The ecological and evolutionary interdependence between web architecture and web silk spun by orb web weaving spiders. *Biol. J. Linn. Soc.* **1987**, *30*, 135–162. [CrossRef]

102. Ulrich, J.C. Wind effects of web structure of grass spider *Agelenopsis actuosa*. *Arachnology* **2021**, *18*, 993–997. [CrossRef]

103. Mulder, T.; Wilkins, L.; Mortimer, B.; Vollrath, F. Dynamic environments do not appear to constrain spider web building behaviour. *Sci. Nat.* **2021**, *108*, 20. [CrossRef] [PubMed]

104. Gunnarsson, B. Fractal dimension of plants and body size distribution in spiders. *Fun. Ecol.* **1992**, *6*, 636–641. [CrossRef]

105. Wolff, J.O.; Little, D.; Herberstein, M.E. Limits of piriform silk adhesion-similar effects of substrate surface polarity on silk anchor performance in two spider species with disparate microhabitat use. *Sci. Nat.* **2020**, *107*, 31. [CrossRef] [PubMed]

106. Hesselberg, T. Ontogenetic changes in web design in two orb-web spiders. *Ethology* **2010**, *116*, 535–545. [CrossRef]

107. Hesselberg, T. The mechanism behind plasticity of web-building behavior in an orb spider facing spatial constraints. *J. Arachnol.* **2014**, *42*, 311–314. [CrossRef]

108. Foelix, R. *Biology of Spiders*, 1st ed.; Oxford University Press: Oxford, UK, 2011.

109. Morley, E.L.; Robert, D. Electric fields elicit ballooning in spiders. *Curr. Biol.* **2018**, *28*, 2324–2330.e2. [CrossRef]

110. Cho, M. Aerodynamics and the role of the earth's electric field in the spiders' ballooning flight. *J. Comp. Physiol. A* **2021**, *207*, 219–236. [CrossRef]

111. Duffey, E. Aerial dispersal in a known spider population. *J. Anim. Ecol.* **1956**, *25*, 85–111. [CrossRef]

112. Lee, V.M.; Kuntner, M.; Li, D. Ballooning behavior in the golden orbweb spider *Nephila pilipes* (Araneae: Nephilidae). *Front. Ecol. Evol.* **2015**, *3*, 2. [CrossRef]
113. Weyman, G.S.; Sunderland, K.D.; Jepson, P.C. A review of the evolution and mechanisms of ballooning by spiders inhabiting arable farmland. *Ethol. Ecol. Evol.* **2002**, *14*, 307–326. [CrossRef]
114. Mowery, M.A.; Lubin, Y.; Segoli, M. Invasive brown widow spiders disperse aerially under a broad range of environmental conditions. *Ethology* **2022**, *128*, 564–571. [CrossRef]
115. Larrivée, M.; Buddle, C.M. Diversity of canopy and understorey spiders in north-temperate hardwood forests. *Agric. For. Entomol.* **2009**, *11*, 225–237. [CrossRef]
116. Ruhland, F.; Chiara, V.; Trabalon, M. Age and egg-sac loss determine maternal behaviour and locomotor activity of wolf spiders (Araneae, Lycosidae). *Behav. Process.* **2016**, *132*, 57–65. [CrossRef] [PubMed]
117. Sergi, C.; Schlais, A.; Marshall, M.; Rodríguez, R.L. Western black widow spiders (*Latrodectus hesperus*) remember prey capture location and size, but only alter behavior for prey caught at particular sites. *Ethology* **2022**, *128*, 707–714. [CrossRef]
118. Rodríguez, R.L.; Gloudeman, M.D. Estimating the repeatability of memories of captured prey formed by *Frontinella communis* spiders (Araneae: Linyphiidae). *Anim. Cogn.* **2011**, *14*, 675–682. [CrossRef]
119. Nørgaard, T.; Henschel, J.R.; Wehner, R. Long-distance navigation in the wandering desert spider *Leucorchestris arenicola*: Can the slope of the dune surface provide a compass cue? *J. Comp. Physiol. A* **2003**, *189*, 801–809. [CrossRef]
120. Gaffin, D.D.; Curry, C.M. Arachnid navigation—A review of classic and emerging models. *J. Arachnol.* **2020**, *48*, 1–25. [CrossRef]
121. Wherry, T.; Elwood, R.W. Relocation, reproduction and remaining alive in an orb-web spider. *J. Zool.* **2009**, *279*, 57–63. [CrossRef]
122. Gillespie, R. The mechanism of habitat selection in the long-jawed orb-weaving spider *Tetragnatha elongata* (Araneae, Tetragnathidae). *J. Arachnol.* **1987**, *15*, 81–90.
123. Beyer, M.; Mangliers, J.; Tuni, C. Silk-borne chemicals of spider nuptial gifts elicit female gift acceptance. *Biol. Lett.* **2021**, *17*, 20210386. [CrossRef]
124. Gerbaulet, M.; Möllerke, A.; Weiss, K.; Chinta, S.; Schneider, J.M.; Schulz, S. Identification of cuticular and web lipids of the spider *Argiope bruennichi*. *J. Chem. Ecol.* **2022**, *48*, 244–262. [CrossRef]
125. Fischer, A.; Lee, Y.; Dong, T.; Gries, G. Know your foe: Synanthropic spiders are deterred by semiochemicals of European fire ants. *R. Soc. Open Sci.* **2021**, *8*, 210279. [CrossRef] [PubMed]
126. Rypstra, A.L.; Schmidt, J.M.; Reif, B.D.; DeVito, J.; Persons, M.H. Tradeoffs involved in site selection and foraging in a wolf spider: Effects of substrate structure and predation risk. *Oikos* **2007**, *116*, 853–863. [CrossRef]
127. Penfold, S.; Dayananda, B.; Webb, J.K. Chemical cues influence retreat-site selection by flat rock spiders. *Behaviour* **2017**, *154*, 149–161. [CrossRef]
128. Dettner, K.; Liepert, C. Chemical mimicry and camouflage. *Annu. Rev. Entomol.* **1994**, *39*, 129–154. [CrossRef]
129. Allan, R.A.; Capon, R.J.; Brown, W.V.; Elgar, M.A. Mimicry of host cuticular hydrocarbons by salticid spider *Cosmophasis bitaeniata* that preys on larvae of tree ants Oecophylla smaragdina. *J. Chem. Ecol.* **2002**, *28*, 835–848. [CrossRef]
130. Elgar, M.A.; Allan, R.A. Predatory spider mimics acquire colony-specific cuticular hydrocarbons from their ant model prey. *Naturwissenschaften* **2004**, *91*, 143–147. [CrossRef]
131. Cerveira, A.M.; Jackson, R.R. Inter-population variation and phenotypic plasticity in kairomone use by a poly-specialist spider-eating predator. *J. Ethol.* **2022**, *40*, 37–48. [CrossRef]
132. Pekár, S. Ant-mimicking spider actively selects its mimetic model (Araneae: Gnaphosidae; Hymenoptera: Formicidae). *Myrmecol. News.* **2020**, *30*, 131–137. [CrossRef]
133. Johnson, J.C.; Trubl, P.; Blackmore, V.; Miles, L. Male black widows court well-fed females more than starved females: Silken cues indicate sexual cannibalism risk. *Anim. Behav.* **2011**, *82*, 383–390. [CrossRef]
134. Junker, R.R.; Bretscher, S.; Dötterl, S.; Blüthgen, N. Phytochemical cues affect hunting-site choices of a nursery web spider (*Pisaura mirabilis*) but not a crab spider (*Misumena vatia*). *J. Arachnol.* **2011**, *39*, 113–117. [CrossRef]
135. Heiling, A.M.; Cheng, K.; Herberstein, M.E. Exploitation of floral signals by crab spiders (*Thomisus spectabilis*, Thomisidae). *Behav. Ecol.* **2004**, *15*, 321–326. [CrossRef]
136. Tedore, C.; Johnsen, S. Immunological dependence of plant-dwelling animals on the medicinal properties of their plant substrates: A preliminary test of a novel evolutionary hypothesis. *Arthropod. Plant Interact.* **2015**, *9*, 437–446. [CrossRef]
137. Li, Z.; Su, Q.; Zhao, Y.; Yun, Y.; Peng, Y. Volatile scents preference of crab spiders (*Ebrechtella tricuspidata*) for host flowers. ResearchSquare PPR482028. *arXiv* **2022**. [CrossRef]
138. Blamires, S.J. Plasticity in extended phenotypes: Orb web architectural responses to variations in prey parameters. *J. Exp. Biol.* **2010**, *213*, 3207–3212. [CrossRef]
139. Haberkern, A.M.; Fernandez-Fournier, P.; Avilés, L. Spinning in the rain: Interactions between spider web morphology and microhabitat use. *Biotropica* **2020**, *52*, 480–487. [CrossRef]
140. Hironaka, Y.; Abé, H. Nesting habits of the Japanese foliage spider, *Cheiracanthium japonicum* (Araneae: Miturgidae): Host plant preference based on the physical traits of plant leaves. *J. Nat. Hist.* **2012**, *46*, 2665–2676. [CrossRef]
141. Itescu, Y. Are island-like systems biologically similar to islands? a review of the evidence. *Ecography* **2018**, *42*, 1298–1314. [CrossRef]
142. Slatkin, M. Gene flow and the geographic structure of natural populations. *Science* **1987**, *236*, 787–792. [CrossRef]

143. Lee, J.W.; Jiang, L.; Su, Y.C.; Tso, I.M. Is Central mountain range a geographic barrier to the giant wood spider *Nephila pilipes* (Araneae: Tetragnathidae) in Taiwan? a population genetic approach. *Zool. Stud.* **2004**, *43*, 112–122.
144. Bell, J.R.; Bohan, D.A.; Shaw, E.M.; Weyman, G.S. Ballooning dispersal using silk: World fauna, phylogenies, genetics and models. *Bull. Entomol. Res.* **2005**, *95*, 69–114. [CrossRef] [PubMed]
145. Amador-Vargas, S.; Orribarra, V.S.; Portugal-Loayza, A.; Fernández-Marín, H. Association patterns of swollen-thorn acacias with three ant species and other organisms in a dry forest of Panama. *Biotropica* **2021**, *53*, 560–566. [CrossRef]
146. Coronado-Rivera, J.; Del Valle, M.S.; Amador-Vargas, S. True bugs living on ant-defended acacias: Evasion strategies and ant species preferences, in Costa Rica and Panama. *Rev. Biol. Trop.* **2020**, *68*, 415–425. [CrossRef]

Article

Validation of a Novel Stereo Vibrometry Technique for Spiderweb Signal Analysis

Nathan Justus [1,*], Rodrigo Krugner [2] and Ross L. Hatton [1]

[1] Laboratory for Robotics and Applied Mechanics, School of Mechanical Industrial and Manufacturing Engineering, Oregon State University, 204 Rogers Hall, Corvallis, OR 97331, USA; ross.hatton@oregonstate.edu

[2] United States Department of Agriculture, Agricultural Research Service, 9611 S. Riverbend Ave, Parlier, CA 93648, USA; rodrigo.krugner@usda.gov

*   Correspondence: justusn@oregonstate.edu

Simple Summary: Spiders often use their webs as sensory mechanisms, obtaining from them such information as the location of prey, the presence of rival spiders, and the characteristics of potential mates. Examining how this information is transmitted through the web and received by spiders is a promising biological area of research that could provide insight into a spider's world and lead to new technologies that leverage these discoveries. In this paper, we develop a novel noncontact technique using two video cameras that is capable of analyzing vibrational signals transmitted through spiderwebs and validate this technique against the current standard of laser Doppler vibrometry. By combining the principles of stereo vision and video vibrometry, we can automatically extract three-dimensional vibrational information at any point in the spiderweb across time, and study how these signals propagate through the web. We show that this technique produces results comparable to those of standard laser vibrometry.

Abstract: From courtship rituals, to prey identification, to displays of rivalry, a spider's web vibrates with a symphony of information. Examining the modality of information being transmitted and how spiders interact with this information could lead to new understanding how spiders perceive the world around them through their webs, and new biological and engineering techniques that leverage this understanding. Spiders interact with their webs through a variety of body motions, including abdominal tremors, bounces, and limb jerks along threads of the web. These signals often create a large enough visual signature that the web vibrations can be analyzed using video vibrometry on high-speed video of the communication exchange. Using video vibrometry to examine these signals has numerous benefits over the conventional method of laser vibrometry, such as the ability to analyze three-dimensional vibrations and the ability to take measurements from anywhere in the web, including directly from the body of the spider itself. In this study, we developed a method of three-dimensional vibration analysis that combines video vibrometry with stereo vision, and verified this method against laser vibrometry on a black widow spiderweb that was experiencing rivalry signals from two female spiders.

Keywords: spiderweb vibrometry; video vibrometry; black widow

Citation: Justus, N.; Krugner, R.; Hatton, R.L. Validation of a Novel Stereo Vibrometry Technique for Spiderweb Signal Analysis. *Insects* **2022**, *13*, 310. https://doi.org/10.3390/insects13040310

Academic Editors: Dumas Gálvez and Thomas Hesselberg

Received: 2 February 2022
Accepted: 16 March 2022
Published: 22 March 2022

Publisher's Note: MDPI stays neutral with regard to jurisdictional claims in published maps and institutional affiliations.

1. Introduction

A spider's web communicates a vast amount of information to its owner, in the form of vibrations that thrum through the structure. Although we are familiar with the ability of spiders to use web vibrations to identify and sense the positions of prey, intruders, and potential mates, the exact mechanisms behind how this information is transmitted remain a mystery, even for the simplest case of the planar orb web [1–5]. For more complex three-dimensional webs, such as those built by the western black widow (*Latrodectus hesperus,*

Chamberlin and Ivie 1935), understanding how a spider might perceive these signals is an even more daunting task. To advance this quest of understanding spiderweb vibrations, we must first develop a reliable and flexible method for experimentally determining the vibrations present in a web that might be sensed by the spider. Recent advances in the field of image processing and computer vision allow for the recovery of motion signals through the analysis of high-speed video, which is the technique we used in this paper to examine and validate visual signatures of rivalry signals in the webs of our model species, *L. hesperus* [6–8].

The current state-of-the-art method for the vibration analysis of spider-related signals is laser Doppler vibrometry, which has commonly been used to examine the webs of *L. hesperus* [9–13] and both three-dimensional webs and planar orb webs built by other species [2–5,14,15]. A typical experiment utilizing a laser vibrometer involves recording vibrational data using one or more laser vibrometers aligned against the web structure or against the body of a stationary spider while the web is undergoing excitation from some signal, typically a shaker that can oscillate the web at a chosen range of frequencies. The insights gained by this spiderweb vibrometry are diverse. Previous studies have examined such things as the signal attenuation in webs for the different vibrational modes to hypothesize which propagation modalities in the web might carry the most consistently valuable information for a spider [2–4,10], the speed of sound and tension in webs to hypothesize how a spider might perform prey localization [4,14] or tune its web to create the most beneficial acoustic characteristics [15,16], and the vibratory characteristics of prey and mate signals that correlate to whether the owner of the web illicits a predatory or a courtship response [9,10].

Using laser vibrometry to analyze spiderwebs has a number of significant challenges: Although it is possible to align a laser vibrometer's beam against a single strand or junction of the web, vibrations often cause the strand to leave the narrow focus of the laser, making this method of data collection unstable [4]. Additionally, *L. hesperus* are often fairly mobile during communication events, eliminating the possibility of aligning the laser against the spider itself, as has been done in former studies where the spider remained stationary in the center [2–4]. Previous studies using laser vibrometers have attempted to counteract this problem by aligning the laser against foreign objects such as squares of reflective tape or fly wings suspended in the web, although this has the potential to change the structure of the collected signals and can be a painstaking process [4,10,12,13]. Another limitation of laser vibrometry is the single-point nature and one-dimensionality of its data. Web vibrations can occur in transverse, longitudinal, and lateral modalities, and the type and quantity of information transmitted by each modality are still unclear [2,4,5,10]. Previous studies have responded to the one-dimensionality of laser vibrometry data by either limiting their investigation to only one of these modalities [12,13], or by performing multiple laser vibrometer recordings at different angles and relying on the planar structure of the web to inform the modality of vibrational data being recorded [2,4,5,10]. Although multiple laser vibrometers can be deployed to increase the information gained per experiment, the capability to compare vibrations across the entirety of the web and for each of the vibrational modes for the same experiment requires a new method of vibrometry.

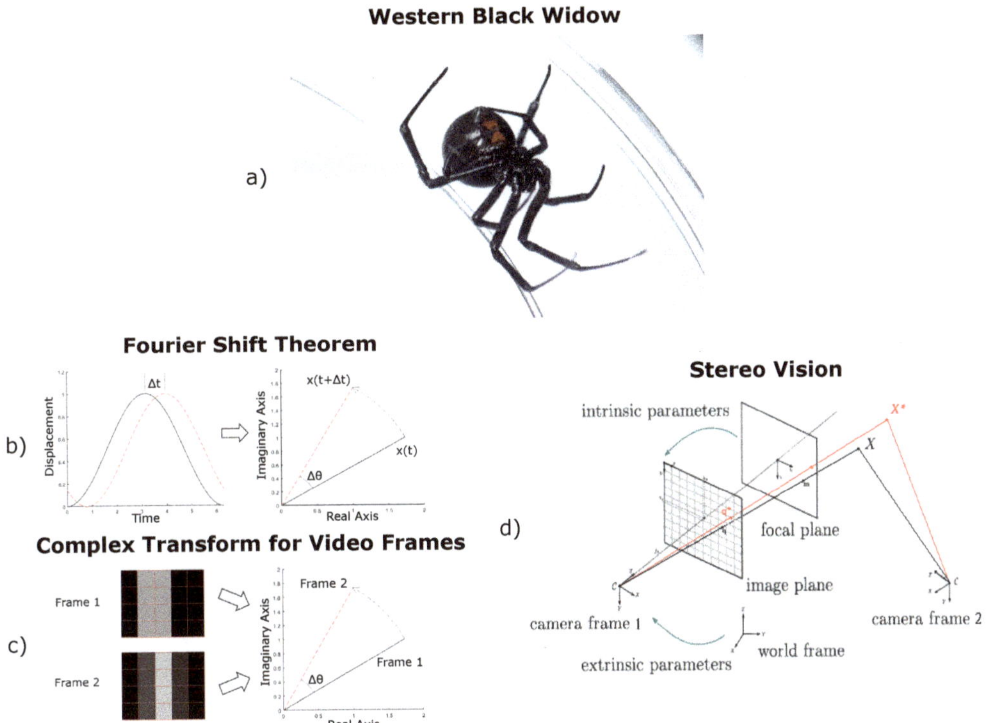

Figure 1. (**a**) Western black widow—*Latrodectus hesperus*. (**b**) Phase shifts from complex transforms: by the Fourier shift theorem, a spatial time-shift of a signal corresponds to a phase-shift in the complex domain. (**c**) Similarly, texture shifts in images correspond to phase-shifts in the complex domain in video vibrometry. Even if a feature moves less than a pixel, this motion registers as changes in pixel brightness values that can be used to automatically estimate motion velocity. (**d**) Pinhole camera model with augmented object position from video vibrometry in red [17]. An image is projected onto a focal plane in the camera's coordinate frame using internal parameters, such as pixel resolution and focal distance, and then the focal plane can be placed in the world frame using external parameters, such as the camera's position and orientation. A line is drawn from the camera focal point to the object in the focal plane. This line passes through the object in 3D space. If pixel movement values are known, this line can be shifted to get an augmented 3D position.

A high-speed video taken of the web as a whole can be used to analyze vibrations visually present in the web or spider without the need for careful alignment of the measurement instrument. Using phase-based video vibrometry on cropped subsections of this video allows for vibrometry to be performed in specific locations without the need for pixel-tracking of specific features, so long as there is sufficient degree of pixel value variation, or "texture" in the chosen analysis region. As this technique extracts information from changes in pixel brightness as texture gradually shifts from one pixel to the next, video vibrometry can give displacement resolutions as small as tiny fractions of a single pixel. The method is sensitive enough that it has been shown capable of reproducing intelligible human speech through analysis of high-framerate video of objects near a person speaking [7]. Although this noncontact technology has been previously used to examine biological phenomena such as microsaccades in the human eye [18,19] and human pulse rates through minute changes in skin color [20], it has yet to reach widespread use in the field of biology. In addition to the convenience of being able to analyze information across the entirety of a spiderweb without regard to specific sample location, video vibrometry provides vibration measurements in both the vertical and horizontal axes of the video, compared to the single depth axis of information collected using a laser [6]. This property makes it possible to apply the principles of stereo vision to video vibrometry, and combine information from two simultaneous videos of the spiderweb to analyze vibrations happening in all three spatial dimensions and at multiple points across time. This is of particular importance to black widow webs, which are highly three-dimensional mesh structures and have multiple vibrational modes in each dimension.

In this study, we developed a novel technique of three-dimensional vibration analysis by combining stereo vision and phase-based video vibrometry, and then applied this technique to extract three-dimensional vibration information from a black widow spiderweb during female–female displays of rivalry. We first describe the technique in detail, and then verify this method by comparing results from stereo vibrometry with information extracted with a laser vibrometer from a paper cube suspended in the web. We then discuss information that we can gain from stereo vibrometry that would be difficult to collect with a laser.

2. Background

This work combines two areas of research: video vibrometry and stereo vision.

2.1. Video Vibrometry

Video vibrometry extracts approximations of pixel velocity values from shifts in pixel brightness values between two adjacent video frames in regions of large pixel value variation without the need for the selection and tracking of specific features. Repeating this process for sets of adjacent frames in the video gives estimates of local motion in the horizontal and vertical coordinates of the video over the length of the examined time. The principles of phase-based video vibrometry lie in the application of the complex-valued pyramid transform, which is constructed by repeatedly applying a complex filter across multiple orientations and spatial scales [6]. This transform moves information from the real-real time domain to the real-imaginary complex domain, which has been shown to be a more reliable format for signal analysis [6,21,22]. Just as a spatially translated signal results in a change in phase in the complex domain through the Fourier shift theorem, spatial translation of pixel intensity values in subsequent frames of a video result in changes in phase through the complex-valued pyramid transform. This concept is illustrated in Figure 1b,c. Complex filters have been used for many computer vision tasks, including image orientation analysis and edge detection [22]. These pyramids are constructed for consecutive frames, and phase differences in these pyramids correspond to spatial shifts in texture.

2.2. Stereo Vision

Two images from different cameras can be combined by leveraging the pinhole camera model, shown in Figure 1d [23]. First, the image is projected from the image plane onto a focal plane in the camera coordinate frame at a fixed distance in front of the origin. This transformation of the image is performed using the camera's intrinsic properties: the resolution of the image, the location of the center pixel, and the focal length of the camera, which determines the millimeters per pixel in the focal plane. The focal plane projection is then moved to the world frame using the camera's extrinsic properties: how the camera is positioned in the world frame and the pan/tilt rotation of the camera. The line connecting the camera's focal point to the representation of an object in the transformed image in the camera's focal plane will pass through that object in 3D space. This process is repeated for the second camera, and the intersection of these two lines is the approximate position of that object in 3D space.

3. Materials and Methods

To conduct our experiment, we first positioned a transparent box containing a female *L. hesperus* specimen and web on a vibration isolation table, along with two cameras and a laser vibrometer. A second female was placed inside the box on the web, and data from the ensuing confrontation were captured from the three instruments. Calibration of the cameras was conducted using approximate measurements taken of their respective positions relative to the spiderweb, and then refined using known measurements from the videos. The data from the communication exchange of web-jerk events from the web owner captured with the cameras were then processed using video vibrometry.

3.1. Experimental Setup

For the experiment, two Chronos 1.4 high-speed monochrome cameras (Kron Technologies Inc., Burnaby, BC, Canada) were rigidly attached to a vibration isolation table such that the only possible allowed movement of the cameras relative to the table were pan and tilt rotations. These cameras were both connected to the same trigger, which signalled capture of the synchronized videos. The cameras were set to record 16 s of video filmed at a framerate of 1000 fps and a resolution of 600 by 800 pixels.

In the center of the vibration isolation table was placed a 30 cm transparent cube in which a female *L. hesperus* had constructed a web. On the front of the cage we placed a 2.54 cm square grid for use in camera calibration. The web was lit from above to provide good contrast of spiderweb features against the dark cage background. Measurements were taken with a tape measure to determine the horizontal position of each camera with respect to the front face of the cage. The maximum angle of the cameras with respect to each other was constrained by visibility into the cage, which was opaque on the sides.

A PDV-100 laser vibrometer (Polytec Inc., Irvine, CA, USA) was positioned between the cameras, and measurements were taken to determine its position relative to them. Days before the experiment, small paper cubes were sprinkled into the spider's web, and the spider was given time to cut down some of the cubes. The laser vibrometer was aligned against one of the remaining cubes.

An "intruder" *L. hesperus* was placed on the web while the spider who built the web was in the retreat. The laser vibrometer was set to record the entire interaction, and the cameras captured 16-second bursts of actions whenever the experiment operator determined a communication exchange was occurring. Video pairs with large amounts of activity were saved and later used for calibration and analysis.

3.2. Camera Calibration

Camera calibration can be performed using any software that minimizes error in predictions for camera pose, camera focal length, and camera pixel size using given calibration points in each video. The specific methodology employed for camera calibration in this work is described in this section.

The world coordinate frame axis directions were defined relative to the front face of the spider cage. Positive X pointed to the right on the front face, positive Y pointed up, and positive Z pointed towards the cameras away from the cage. The positions of the cameras in the XZ plane were assumed known from the measurements taken during experimental setup, and the resolution of the video and location of the center pixels used for the intrinsic transformation were known from the camera settings. The remaining unknown parameters were the pan and tilt angles of each camera, the Y translation of the cameras relative to the spider cage, and the focal distances of the cameras.

To eliminate the need for knowledge of camera Y displacement relative to the vibration isolation table, the origin was chosen as the point halfway between the walls of the cage lying on the horizontal epipolar line of the video. The horizontal epipolar line was found by locating the intersection of the eight lines in each video that are parallel to the XZ plane (the top and bottom of each cork triplet, and top and bottom of each horizontal slit in the walls of the cage, seen in Figure 2). As the only allowed rotations of each camera are pan and tilt, the horizontal epipolar line represents the XZ plane that contains the focal points for both cameras, and Y translation can be considered zero.

To calculate the remaining unknowns of pan angle, tilt angle, and focal distance, calibration points were selected that were known to lie on the XY plane. As the dimensions of the cage were known, the points on the horizontal epipolar line that intersect with the cage walls have known positions. For each 2.54 cm square grid taped to the front of the cage, the four corners have known distances relative to each other.

An initial guess for the pan angle, tilt angle, and the focal distance were taken, and the discretized parameter-space around this point was exhaustively searched for the configuration that minimized the summed squared error of the origin, the cage wall points, and the distances between the points on the calibration squares for each transformation. Resulting errors between the known locations of the calibration points and the estimated positions of calibration points projected onto the front of the cage using the minimum-error configuration were all less than half of a millimeter. This minimum-error prediction of the pan/tilt angles and the focal distance along with the known camera translation and resolution comprised the final transforms used to analyze stereo data of the spiderwebs.

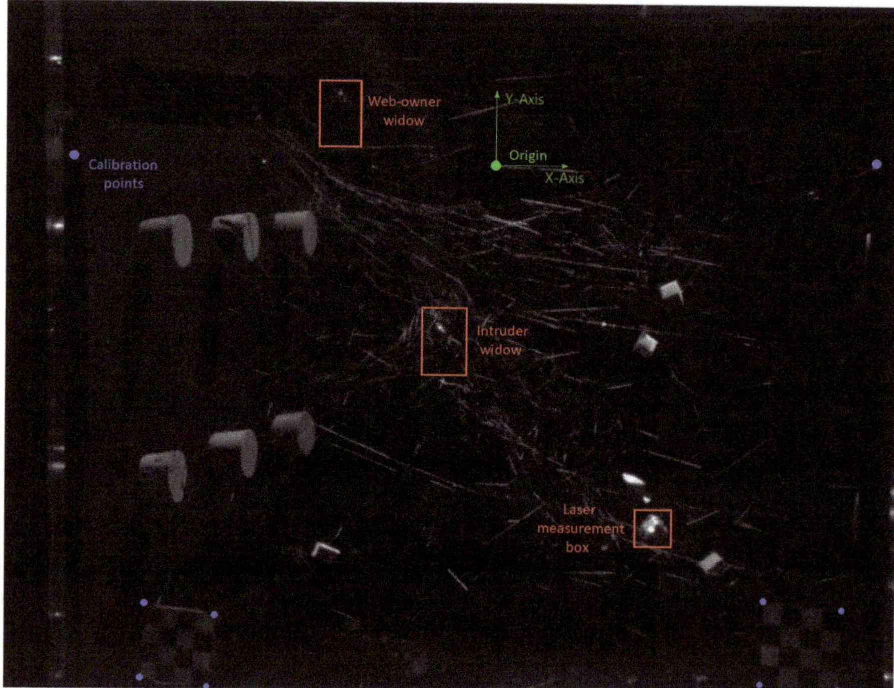

Figure 2. A frame from the right camera video used in the calibration process: The green point represents the chosen origin along with coordinate axes, and blue points represent coordinates with known distances to the origin or between each other. Regions highlighted in red were cropped in order to obtain local motion signals for that area of the web. A strength of the video vibrometry technique is that although the spiders are barely visible in this frame, local variations in pixel intensity over the course of the video are sufficient to predict vibrational motion for the spiders without the need for pixel tracking of specific spider features.

3.3. Stereo Analysis

In order to extract the stereo vibration information from the videos, each video was first cropped down to the region to be analyzed. This was done for both the resident and intruder spider and for the paper cube that the laser was aligned against.

Complex-steerable pyramids were constructed for each cropped video using the complex filter taps outlined in [22], which have been used previously for similar work [6,21]. For each frame of the video, the layers of the pyramid are constructed by downsampling the base image of the previous layer by a factor of two, then applying a complex filter pair to the downsampled base image for both the horizontal and vertical direction in the video. "Phase changes", which correspond to shifts in pixel intensity between two frames as objects move relative to the camera, were extracted from the pyramids for each pixel across both videos and the phase signals were denoised: first spatially using an amplitude-weighted blur to squash noise in textureless regions of the video, and then temporally using butterworth mid-pass filters built to pass signals from 5 to 100 Hz. The denoised phase velocity signals were integrated to get displacement estimates for each pixel, and the displacement signals were averaged over the entirety of the cropped region to give local motion estimates in camera frame X and Y in units of pixels for each region of interest. These video vibrometry operations were performed with the aid of a Matlab GUI written for this purpose, available on github https://github.com/NathanJustus/VideoVibrometry_MatlabApp (accessed on 15 March 2022).

Stereo data composition was achieved by first finding initial pixel coordinates representing the spider or cube being analyzed in the first frame of each video. This task was performed by hand for each camera. These points were then projected onto that camera's focal plane in the world coordinate frame using the intrinsic and extrinsic transformation parameters found during camera calibration, and constructing the line connecting that camera's focal point to the point to be examined in the focal plane. The point that minimized the distance to the corresponding line from each camera served as our guess for the object's initial position in space.

The initial pixel position of the object in each video was then augmented using the pixel displacement estimates calculated using video vibrometry, and the transformation process was repeated to obtain the new predicted position. Carried out for both videos, this produced a 3D local motion estimate across time for the region being examined.

Finally, for comparison with the laser vibrometer, the 3D displacement data were projected onto the line connecting the predicted coordinates of the paper cube the laser was aligned against to the measured position of the laser vibrometer relative to the cameras. Stereo data were aligned to the laser data manually by examining the time signatures of major signalling events.

4. Results

To examine the results of this experiment, we first validate measurements of the signals created by the *L. hesperus* and captured with stereo vibrometry against the measurements of the signals captured with laser vibrometry. Once our confidence in the method is confirmed, we examine data collected with stereo vibrometry for insights that cannot be attained through similar measurements taken with laser vibrometry.

4.1. Signal Pattern Analysis of Paper Box

From the laser vibrometry web displacement estimates illustrated in the top portion of Figure 3a, it is clear that the signals were composed of two sets of three individual events. The first three events (from 0 to 5 s) are visible only in the laser vibrometry signal and are much smaller. These events lie within the noise of the stereo data from the paper box. For the laser, the noise in regions with no recognizable signal is around 1 mm, and the noise of the stereo method in the same region is approximately 2 mm. Most of the noise present in the stereo vibrometry data is likely due to stereo calibration error and to the fact that the vibrations were projected onto the axis of laser vibrometer measurements, which is one of the hardest axes to measure. As motion in this axis tends to move objects towards and away from each camera rather than side to side or up and down, this motion is fairly difficult to detect using stereo vision. The raw video vibrometry displacement data have an ambient noise of approximately one hundredth of a pixel, making it much less noisy to measure signals in other axes. This noise could likely be improved by mounting the cameras further apart from each other or by using better software specifically designed for the general problem of camera calibration for stereo vision.

The latter three signals (from 6 to 16 s) were more intense and were picked up fairly equally by both stereo and laser vibrometry. Both methods estimate peak displacement amplitudes of 20 to 30 mm, but the stereo vibrometry signal prediction decays slightly faster than the laser vibrometry prediction. These are rather large-amplitude vibrations. In general, stereo vibrometry will be most useful when analyzing motions that are large enough to cause significant changes in pixel brightness (likely on the order of a mm or so in the case of spiderwebs depending on camera choice and experimental setup) but are also small enough that they do not cause relevant objects to leave the frame of the video.

It is also worthy of note that the laser vibrometer was oriented approximately with the Z axis. The cameras each have a rotation angle of approximately 15 degrees with respect to the Z axis, giving them individually very little information about vibrations in this axis. However, by combining their information using this stereo technique, vibrations in this most difficult axis to measure can produce results comparable to those of a laser vibrometer.

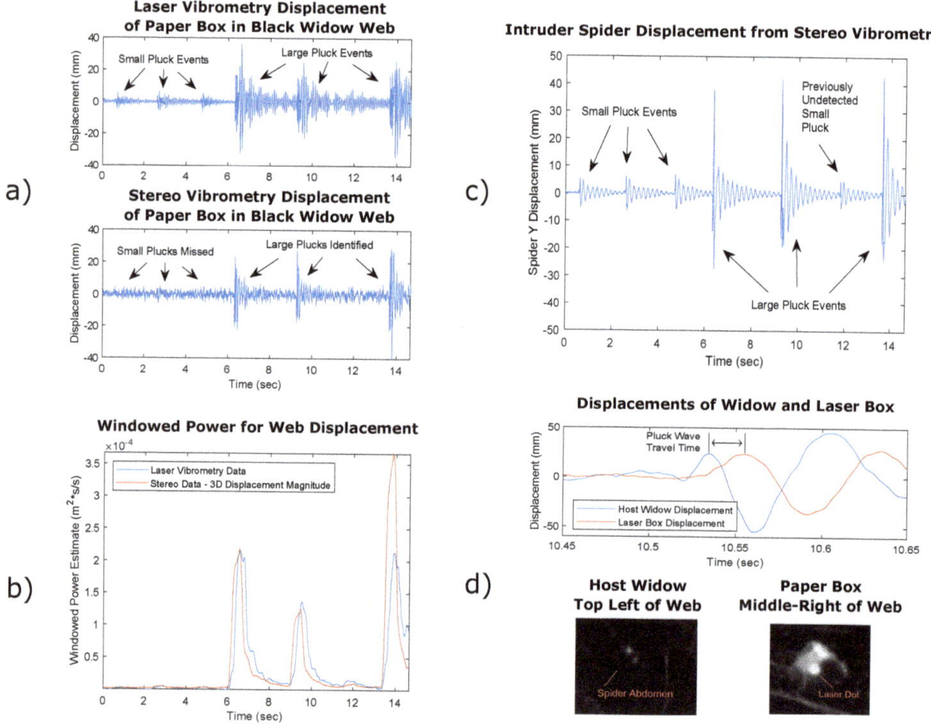

Figure 3. (**a**) Stereo vibrometry comparison with laser vibrometry: The stereo vibrometry data of a paper cube suspended in the web projected to the laser vibrometry axis correctly identify three large web jerk events but miss small signals detected by the laser vibrometer. (**b**) Windowed power estimates of spiderweb signals: Time-domain power can be used to detect large signals for both laser vibrometry and stereo vibrometry. (**c**) Intruder spider displacement from stereo vibrometry: Using stereo vibrometry on a spider itself rather than a paper cube generates much cleaner signals. (**d**) Calculation of the wave speed of the web signal: Video subregions are analyzed using stereo vibrometry, and the time delay between signal peaks is used to calculate the wave travel time. Wave speed is calculated using wave travel time and 3D position estimates from stereo vision.

4.2. Signal Power Analysis of Paper Box

Both the laser vibrometry data and stereo vibrometry data were further analyzed by applying a filter that estimates time-domain signal power. For stereo vibrometry, this analysis was performed on the total 3D displacement magnitude predicted of the paper cube. The results are illustrated in Figure 3b.

The time-domain power estimate comparison with laser vibrometry shows that the 3D displacement magnitude from stereo vision correctly estimates times of peak power. This verifies our confidence that stereo vibrometry supplies accurate measurements of signals in the web. This time domain analysis of stereo vibrometry data is also a technique that could be useful for future work analyzing the *L. hesperus* rivalry displays, as it would be feasible to use this result to automate the extraction of large black widow rivalry signal patterns.

4.3. Stereo Vibrometry Analysis of Intruder Spider

Having verified the stereo vibrometry technique against laser vibrometry, we can use this process to take measurements not possible with the laser vibrometer. When analyzing spider rivalry signals, it will likely be most important to measure the signal felt by the spiders themselves. This measurement cannot be taken with the laser, as the spider shifts around in the web during the rivalry displays, and the laser must be aligned against a stationary point. Video vibrometry, however, allows these signals to be measured during times in the video where the spider remains stationary, even if it has shifted after camera setup. The y-axis stereo vibration signal for the intruder spider is shown in Figure 3c.

In the displacement signal from the intruder spider, the smaller signals that were hidden in the paper cube stereo data are now clearly visible. It is also possible to see a fourth small vibration event between two larger signals at 13 s. This event was hidden by noise in both the laser vibrometry and stereo vibrometry data from the paper cube in Figure 3a. This hidden signal found with stereo vibrometry shows that using vibration signals taken directly from the black widow spiders with stereo vibrometry as opposed to vibration signals from paper cubes sprinkled in the web can reveal important rivalry signalling patterns that would otherwise be missed using conventional techniques. This measurement is only possible using video vibrometry.

As the high mass of the intruder spider relative to the paper cube causes it to vibrate at a lower frequency with a cleaner signal, the vibration of the spider is much easier to read with stereo vibrometry than the signal from the paper cube. Additionally, the texture in the cropped region is dominated by the intruder spider itself rather than individual strands of the web, so the resulting signal has much less noise.

As noted earlier, because the cameras are positioned at an orientation of approximately 15 degrees with respect to the Z axis, reconstructing vibrations aligned with the axis of the laser vibrometer produces fairly noisy results. However, by choosing to analyze the intruder spider vibrations in the Y direction, to which both cameras can capture the full range of motion, the two camera vibration signals act as multiple sensors reading the same motion, producing lower levels of noise in the combined result. This aligns with our understanding of binocular vision in general: it is more difficult to perceive motion coming towards or away from the viewer than it is to perceive motion going side-to-side or up-and-down.

4.4. Web Signal Wave Speed Estimation

Rather than analyzing individual points in space, we can also use stereo vibrometry to characterize the behavior of the web as a whole. For instance, we can use this technique to estimate the wave speed of the signal as it travels through the web. By examining the time difference between the first peak displacements of the three large signal events at the origin of the signal and at the laser-aligned paper cube, we can approximate the time it takes for the signal to travel across the web. By combining this information with our knowledge of the three-dimensional positions of these points in space from stereo vision, we can calculate the wave speed of the signal as it travels through the web. An illustration of this process can be seen in Figure 3d. For the three large signals in the video, the calculated wave speeds were 24.2 m/s, 17.7 m/s, and 16.6 m/s, for an average predicted wave speed of 19.5 m/s. This estimate is fairly low compared to previous attempts to measure low-frequency transverse wave speeds using multiple sensors on the web, which have given results of 67.6 m/s [4] and 109 m/s [14]. Both of these measurements were performed on planar orb webs, so it is unclear how the speed of sound would be affected by a web with a three-dimensional mesh structure. Such measurements have been used in the past to make predictions about how a spider might perform prey localization [4,14], gather information using signal lines while away from the center of the web [15,16], and alter the structure of the web to create better acoustic properties [5,16].

5. Discussion and Future Applications

There are a few considerations that must be kept in mind when deciding to use stereo vibrometry to examine vibrations in spiderwebs or other three-dimensional structures. The first is that the frequency bandwidth of this method depends on the framerate of the camera being used to record the vibrations. The Nyquist frequency (one half of the camera framerate) determines the absolute maximum possible vibration frequency that can be recorded. However, this upper bound can prove quite noisy, and it is more typical to limit analysis to a quarter of the sampling frequency: a 1000 frames-per second recording would typically only be used for frequency analysis up to around 250 Hz, whereas laser vibrometry has a very high sample rate and is typically rated to measure signals up to 2 MHz [24].

Another consideration that must be thought through before employing this technique is that of vibrational signal noise, which will depend on experimental design and the vibrational axis chosen to examine. Although the resolution of phase-based video vibrometry has been shown to be accurate down to a few thousandths of a pixel [6], it is still of much lower resolution than a laser vibrometer, which is rated on the order of nanometers [24]. Performing stereo vision also increases the measurement noise over that of two-dimensional video vibrometry because of camera calibration error and binocular vision effects. A camera can only effectively measure motion that moves features horizontally or vertically in the image plane, and so vibrational information from stereo vibrometry will have noise dependent on the desired measurement axis. In Figure 3a, the stereo vibrations were projected onto the axis of the laser vibrometer, one of the noisiest possible axes, giving a noise of around 2 mm. However, in Figure 3c a better axis is chosen, giving a noise of around 1/3 of a mm.

In general, laser vibrometry will be more specialized at reading vibrations in a single dimension at an individual focus point, giving better frequency ranges and sensitivity, whereas stereo vibrometry allows for more general sample collection throughout the entirety of the web and in all dimensions at the cost of being limited by the frame-rate of the camera and the minimum resolution of a pixel allowed by camera focus and placement of the camera relative to the subject material. When considering experimental design, a camera with maximum feasible resolution should be placed as close as possible to the spiderweb in order to maximize pixel density throughout the web, while simultaneously ensuring that the camera is far enough away that proper focus can be achieved and that the image is sharp. When considering noise in each vibrational axis, the orientation of the two cameras with respect to each other plays an important role. Ideally, the two cameras should be orthogonal. However, this is often not possible because of spider cage design, so it should be acknowledged that when the cameras are closely aligned, it will be more difficult to detect motion data in the direction going towards or away from the camera lenses.

Despite the considerations that must be employed for this technique, the benefits are immense. This technique does not rely on pixel tracking of features but rather autonomous detection of pixel intensity variations between camera frames, allowing for resolutions much lower than a single pixel [7]. As stereo vibrometry gives 3D displacement data, the technique allows for the simultaneous analysis of all of the longitudinal, transverse, and lateral vibration modes along the spiderweb, presuming that these axes of vibration can be intelligently chosen. Another convenience is that video vibrometry can be performed on any subsection of the video that has adequate focus [6–8]. This means that stereo vibrometry allows for simultaneous sampling across the entirety of the spiderweb, allowing for simple investigations into spiderweb characteristics such as signal propagation speed and attenuation rates for each of the different vibrational modalities. Although this study focused specifically on validating stereo vibrometry against laser vibrometry and performing vibrometry on the spider bodies themselves, potential future work could perform vibrometry on nodes in the web mesh itself, giving insights into how spider signals change as they propagate through the web.

For our particular model species of *L. hesperus*, we hope that this technique can provide insights into female–female rivalry signals, which we hypothesize could lead to novel chemical-free signal mimicry techniques in arachnid dispersion that minimize the chances of black widows being collected with table grape clusters during harvest. Similar research into vibrational mating signals of another grapevine pest, the glassy-winged sharpshooter *Homalodisca vitripennis* [25], has shown that these signals can be exploited and mimicked to cause disruptions in mating behavior without the application of chemical pesticides [26]. Due to export concerns, pesticides are currently deployed to minimize accidental widow collection during grape harvest, and non-chemical forms of arachnid dispersion could alleviate undesirable environmental impacts of pesticides that are deployed to kill otherwise beneficial inhabitants of the grapevine ecosystem [27]. Other uses of this technique for spiderwebs include examinations of courtship signals and discovering how spiders might locate their prey.

Although this paper specifically investigates spiderweb signals, we suspect that this technique is also applicable to other fields of biology that are typically inaccessible to study through laser vibrometry, such as the communication of bees and wasps, wing beat patterns of hummingbirds, and even vibrations in thin plant membranes. Any source of vibration that can be captured with a high-speed camera may become subject to biological analysis.

6. Conclusions

In this paper, we described an efficient method for measuring 3D motion for vibration analysis by combining the concepts of stereo vision and video vibrometry. We extracted the local phases from two videos using video vibrometry, used this phase to estimate local motion over time in the camera frame, and then used these motion estimates in a stereo vision model to augment estimates of the 3D position of objects of interest. We then implemented this method to analyze vibrations in female–female *Latrodectus hesperus* displays of rivalry and demonstrated that stereo vibrometry produces results comparable to those of a laser vibrometer. We also discussed measurements that can be taken from the spiderweb using stereo vibrometry that are difficult to achieve using laser vibrometry, such as analyzing vibrations felt by the *L. hesperus* themselves and estimating the wave speed of rivalry signals through the web. We think that this technique can make spiderweb vibrometry more convenient by reducing the experimental burden of laser vibrometry, and that it opens the door to new studies into how signals propagate through the structure of a spiderweb.

Author Contributions: Conceptualization, N.J., R.K. and R.L.H.; Data curation, R.K.; Funding acquisition, R.K. and R.L.H.; Investigation, N.J. and R.K.; Methodology, N.J. and R.K.; Project administration, R.K. and R.L.H.; Resources, R.K.; Software, N.J.; Supervision, R.L.H.; Validation, R.K.; Writing—original draft, N.J.; Writing—review & editing, N.J., R.K. and R.L.H. All authors have read and agreed to the published version of the manuscript.

Funding: This work was funded by the California Table Grape Commission. USDA-ARS appropriated project #2034-22000-012-00D. The work was also supported in part by the National Science Foundation under grant PoLS-1504428.

Data Availability Statement: The video and laser vibrometry files that were used to generate the results in this manuscript can be accessed online through the open science foundation https://osf.io/3xyc5/?view_only=6c3eaf6650224fd7934449bdabb75502 (accessed on 15 March 2022). The MATLAB tools written and used to perform the video vibrometry analysis can be accessed through github https://github.com/NathanJustus/VideoVibrometry_MatlabApp (accessed on 15 March 2022).

Acknowledgments: We thank Theresa de la Torre and Crystal Espindola for technical assistance.

Conflicts of Interest: The authors declare no conflict of interest.

References

1. Kawano, A.; Morassi, A. Can the spider hear the position of the prey? *Mech. Syst. Signal Process.* **2020**, *143*, 106838. [CrossRef]
2. Masters, W.M.; Markl, H. Vibration Signal Transmission in Spider Orb Webs. *Science* **1981**, *213*, 363–365. [CrossRef] [PubMed]
3. Masters, W.M. Vibrations in the orbwebs of Nuctenea sclopetaria (Araneidae). *Behav. Ecol. Sociobiol.* **1984**, *15*, 207–215. [CrossRef]
4. Landolfa, M.; Barth, F. Vibrations in the orb web of the spider Nephila clavipes: Cues for discrimination and orientation. *J. Comp. Physiol. A* **1996**, *179*, 493–508. [CrossRef]
5. Mortimer, B.; Soler, A.; Siviour, C.R.; Vollrath, F. Remote monitoring of vibrational information in spider webs. *Sci. Nat.* **2018**, *105*, 37. [CrossRef] [PubMed]
6. Wadhwa, N.; Rubinstein, M.; Durand, F.; Freeman, W.T. Phase-based Video Motion Processing. *ACM Trans. Graph.* **2013**, *32*, 80:1–80:10. [CrossRef]
7. Davis, A.; Rubinstein, M.; Wadhwa, N.; Mysore, G.J.; Durand, F.; Freeman, W.T. The Visual Microphone: Passive Recovery of Sound from Video. *ACM Trans. Graph.* **2014**, *33*, 79:1–79:10. [CrossRef]
8. Chen, J.G.; Davis, A.; Wadhwa, N.; Durand, F.; Freeman, W.T.; Büyüköztürk, O. Video Camera Based Vibration Measurement for Civil Infrastructure Applications. *J. Infrastruct. Syst.* **2017**, *23*, B4016013. [CrossRef]
9. Vibert, S.; Scott, C.; Gries, G. A meal or a male: the 'whispers' of black widow males do not trigger a predatory response in females. *Front. Zool.* **2014**, *11*, 4. [CrossRef]
10. Vibert, S.; Scott, C.; Gries, G. Vibration transmission through sheet webs of hobo spiders (*Eratigena agrestis*) and tangle webs of western black widow spiders (*Latrodectus hesperus*). *J. Comp. Physiol. A* **2016**, *202*, 749–758. [CrossRef]
11. Mhatre, N.; Sivalinghem, S.; Mason, A.C. Posture controls mechanical tuning in the black widow spider mechanosensory system. *bioRxiv* **2018**. [CrossRef]
12. Sivalinghem, S.; Mason, A.C. Vibratory communication in a black widow spider (*Latrodectus hesperus*): signal structure and signalling mechanisms. *Anim. Behav.* **2021**, *174*, 217–235. [CrossRef]
13. Sivalinghem, S.; Mason, A.C. Function of structured signalling in the black widow spider Latrodectus hesperus. *Anim. Behav.* **2021**, *179*, 279–287. [CrossRef]
14. Mortimer, B.; Gordon, S.D.; Holland, C.; Siviour, C.R.; Vollrath, F.; Windmill, J.F.C. The Speed of Sound in Silk: Linking Material Performance to Biological Function. *Adv. Mater.* **2014**, *26*, 5179–5183. [CrossRef] [PubMed]
15. Mortimer, B.; Soler, A.; Siviour, C.R.; Zaera, R.; Vollrath, F. Tuning the instrument: sonic properties in the spider's web. *J. R. Soc. Interface* **2016**, *13*, 20160341. [CrossRef] [PubMed]
16. Mortimer, B.; Holland, C.; Windmill, J.F.C.; Vollrath, F. Unpicking the signal thread of the sector web spider Zygiella x-notata. *J. R. Soc. Interface* **2015**, *12*, 20150633. [CrossRef]
17. Forsyth, D.; Ponce, J. *Computer Vision, A Modern Approach*; Pearson Education, New York City, NY, USA, 2012.
18. Wu, H.Y.; Rubinstein, M.; Shih, E.; Guttag, J.; Durand, F.; Freeman, W.T. Eulerian Video Magnification for Revealing Subtle Changes in the World. *ACM Trans. Graph.* **2012**, *31*, 1–8. [CrossRef]
19. Rubinstein, M.; Wadhwa, N.; Durand, F.; Freeman, W.T. Revealing Invisible Changes in the World. *Science* **2013**, *339*, 519.
20. Elgharib, M.A.; Hefeeda, M.; Durand, F.; Freeman, W.T. Video magnification in presence of large motions. In Proceedings of the 2015 IEEE Conference on Computer Vision and Pattern Recognition (CVPR), Boston, MA, USA, 7–12 June 2015. [CrossRef]
21. Wadhwa, N.; Rubinstein, M.; Durand, F.; Freeman, W.T. Quaternionic Representation of the Riesz Pyramid for Video Magnification. 2014. Available online: https://www.semanticscholar.org/paper/Quaternionic-Representation-of-the-Riesz-Pyramid-Wadhwa-Rubinstein/31d6d9c988c736008c26184d86a010143d49477d (accessed on 15 March 2022).
22. Freeman, W.T.; Adelson, E.H. The Design and Use of Steerable Filters. *IEEE Trans. Pattern Anal. Mach. Intell.* **1991**, *13*, 891–906. [CrossRef]
23. Murray, D.; Little, J.J. Using Real-Time Stereo Vision for Mobile Robot Navigation. *Auton. Robot.* **2000**, *8*, 161–171. [CrossRef]
24. Johansmann, M.; Siegmund, G.D.; Pineda, M. Targeting the Limits of Laser Doppler Vibrometry. 2005. Available online: https://www.semanticscholar.org/paper/Targeting-the-Limits-of-Laser-Doppler-Vibrometry-Johansmann-Siegmund/fe68e48d8791baa52534b7b2f066f0b0881e11b8 (accessed on 15 March 2022).
25. Nieri, R.; Mazzoni, V.; Gordon, S.D.; Krugner, R. Mating behavior and vibrational mimicry in the glassy-winged sharpshooter, Homalodisca vitripennis. *J. Pest Sci.* **2017**, *90*, 887–899. [CrossRef]

26. Hill, P.S.M.; Lakes-Harlan, R.; Mazzoni, V.; Narins, P.M.; Virant-Doberlet, M.; Wessel, A. *Biotremology: Studying Vibrational Behavior*; Springer Nature: Berlin, Germany, 2019; Chapter 18. [CrossRef]
27. Bentley, W.J. The integrated control concept and its relevance to current integrated pest management in California fresh market grapes. *Pest Manag. Sci.* **2009**, *65*, 1298–1304. [CrossRef] [PubMed]

Article

The Use of Tuning Forks for Studying Behavioural Responses in Orb Web Spiders

Mollie S. Davies [1] and Thomas Hesselberg [1,2,*]

[1] Department of Biological and Medical Sciences, Headington Campus, Oxford Brookes University, Oxford OX3 0BP, UK; mdavies08@qub.ac.uk

[2] Department of Zoology, University of Oxford, 11 Mansfield Road, Oxford OX1 3SZ, UK

* Correspondence: thomas.hesselberg@zoo.ox.ac.uk

Simple Summary: Spiders are common predators found in almost every type of environment, and are used as model organisms in studies ranging from communication and signalling to biochemical studies on their silk. Orb spiders are particularly interesting, as their web provides a cost-effective way to obtain information on their foraging behaviour. However, studies on short-term behaviours including prey capture and escape behaviours are rare and usually take place in artificial settings, such as laboratories. In this study, we tested a simple methodology using tuning forks that can be used consistently and reliably in the field. The two tuning forks are capable of producing attack (440 Hz) and escape (256 Hz) responses from the spiders. We also used a metal wire as a mechanical stimulus for comparison, which as predicted, was less reliable. We demonstrate the usefulness of the methodology by quantitatively investigating how the size of the spider and the size of its web affect predatory and escape response rates in the autumn spider, although no significant effects of either were found. However, our results confirm the ease by which this simple method can be used to conduct behavioural studies of orb spiders in the wild.

Citation: Davies, M.S.; Hesselberg, T. The Use of Tuning Forks for Studying Behavioural Responses in Orb Web Spiders. *Insects* **2022**, *13*, 370. https://doi.org/10.3390/insects13040370

Academic Editor: Vojislava Grbic

Received: 4 March 2022
Accepted: 6 April 2022
Published: 9 April 2022

Publisher's Note: MDPI stays neutral with regard to jurisdictional claims in published maps and institutional affiliations.

Abstract: Spiders and their webs are often used as model organisms to study a wide range of behaviours. However, these behavioural studies are often carried out in the laboratory, and the few field studies usually result in large amounts of video footage and subsequent labour-intensive data analysis. Thus, we aimed to devise a cost- and time-effective method for studying the behaviour of spiders in the field, using the now almost forgotten method of stimulating webs with tuning forks. Our study looked at the viability of using 256 Hz and 440 Hz tuning forks to stimulate, anti-predatory and predatory responses in the orb web spider *Metellina segmentata*, respectively. To assess the consistency of the behaviours produced, we compared these to direct mechanical stimulation with a metal wire. The results suggest that the tuning forks produce relatively consistent behaviours within and between two years in contrast to the metal wire. We furthermore found no significant effects of spider length or web area on spider reaction times. However, we found significant differences in reaction times between escape and prey capture behaviours, and between tuning forks and the wire. Thus, we demonstrated the potential of tuning forks to rapidly generate quantitative data in a field setting.

Keywords: prey capture behaviour; anti-predatory behaviour; tetragnatha; ethogram; vibration

1. Introduction

Spiders are the most well-known, most diverse, and arguably the most interesting of the Arachnida. This has led to their use as model organisms in a wide variety of studies, ranging from animal signalling and communication [1] and invertebrate cognition [2] to biomechanical studies on silk [3]. Those that build webs are particularly interesting as, by using their webs as extended phenotypes, they are capable of detecting and responding to a wide variety of stimuli [4,5]. Orb spiders are the best-studied web-building spiders for a number of reasons; (i) they are abundant and widespread in most ecosystems, (ii) their highly structured two-dimensional webs are easy to quantify in both the laboratory and in the field [6,7], and (iii) their webs show impressive plasticity in their geometrical structure in response to a wide range of physiological and environmental factors [8,9].

Further, the use of spiders in behavioural studies is supported by sufficient documentation of their anti-predatory and predatory behaviours, from families to specific species. The anti-predatory responses of spiders are best summarised by Cloudsley-Thompson [10]. Their predators fall into two main categories, similar-sized arthropods, and vertebrates that are much larger. They employ a range of primary and secondary defences against each predator type [10]. Primary defences reduce the chances of a predator encounter and include living in crevices, beneath bark or within holes (anachoresis), crypsis/protective colouration, and phenological adaptations [10]. Secondary defences are more active, and in orb web spiders, they commonly include flight, thanatosis and rebuff. Flight involves the spider dropping from the web, which becomes more successful when followed by thanatosis and/or crypsis. The behaviour of rebuff was first described by Tolbert [11], during an experiment conducted on *Argiope aurantia* (Lucas, 1833) and *Argiope trifasciata* (Forsskål, 1775), and involves the spiders actively pushing away the stimulus using their front legs. When acting as a predator, spiders rely on tactile and vibratory cues, detected through the surface of a leaf or web silk [12]. The predatory tactics of spiders include active pursuit, sit-and-wait, prey attraction and cautious stalking. However, their dependence on vibratory cues to reveal prey means sit-and-wait predatory approaches are the most common [12]. Orb spiders' prey-capture strategy consists of an initial reaction of turning towards prey caught in the web followed by gathering information on the location and possible size of the prey based on vibrations generated by the struggling prey, which is occasionally supplemented by plucking the threads with its first pair of legs [13,14]. The spider then approaches the prey before attacking it by either biting and/or wrapping it with silk [14]. However, these behaviours are almost exclusively studied under unnatural conditions in the laboratory (examples include [15–17]). Field studies on these naturally rare events remain scarce, and when they do occur, the methods usually involve large quantities of video footage, and subsequently, time-consuming analysis [18,19] (but see [20]). Thus, to obtain an integrated view of spider behaviour in the wild, there is a need to develop an easier and more reliable method. Here, we suggest that one such cost-effective method could be the now almost forgotten practice of stimulating behavioural responses with tuning forks contacting the web thread.

This method takes advantage of the highly geometrical structure of the standard orb web with radii radiating outwards from the hub towards the peripheral frame threads, which enclose the web with sticky spiral threads overlain on the radii [21]. The stronger and stiffer radii and frame threads transmit vibratory information from struggling prey to the spider waiting in the hub of the web, and are therefore the ideal threads to stimulate with tuning forks [22,23]. The use of tuning forks to study spiders has a long history, with Boys' study from 1880 [24] being one of the earliest documented examples. Boys approached orb web garden spiders (species not given, but likely *Araneus diadematus* (Clerck, 1757) in the family Araneidae) with a 440 Hz tuning fork and observed the spiders facing and approaching the tuning fork—a clear demonstration of predatory behaviour [24]. In a more elaborate experiment, Barrows [25] studied the effect of three tuning forks (100 Hz, 487 Hz and an adjustable fork) and an electric vibrator on the orb weaving spider *Larinioides sclopetarius* (Clerck, 1757) in the family Araneidae. It was concluded that the spider orientated itself and moved toward the stimulus when it was vibrating at an appropriate rate and amplitude [25]. It was also determined that the sensory organs used in detecting the vibrating stimulus are likely to be mechanosensitive trichobothria found on the tarsus and metatarsus of the legs [25,26]. Bays [27] conditioned the orb web spider *Araneus diadematus* (Clerck, 1757) in the family Araneidae to respond to different prey options, which were associated with two frequencies, 256 Hz and 523.3 Hz. His experiment revealed that orb web spiders are able to distinguish between different vibrational frequencies, and that they adapt their behaviour based on previous experience [27]. More recently, Justice et al. [28] studied the orb web spider *Argiope florida* (Chamberlin and Ivie, 1944) in the family Araneidae and elicited predatory behaviours using a frequency of 100 Hz. Whilst in an experiment conducted by Nakata and Mori [29], a 440 Hz frequency was used to produce anti-predatory web-building behaviours (webs were more symmetrical in the presence of predator cues) in *Cyclosa argenteoalba* (Bösenberg and Strand, 1906) and *Eriophora sagana* (Bösenberg and Strand, 1906), both in the family Araneidae.

This historic use of tuning forks demonstrates that they are useful for studying the predatory and anti-predatory behaviours of orb web spiders. However, the majority of the studies mentioned above were conducted in a laboratory setting, and only medium-to-large-sized araneid orb spiders were used. This, combined with the variability in the behavioural responses generated by different species and different tuning forks, suggests there is a need for more rigorous and quantitative studies on a wide range of species. Therefore, in this study, we investigated whether two different tuning forks (256 Hz and 440 Hz) could be used to produce anti-predatory and prey-capture behavioural responses in the orb web spider *Metellina segmentata* (Clerck, 1757) from the family Tetragnathidae, in the field. Additionally, we compared the consistency of the behavioural responses to tuning forks to those from a mechanical stimulation of the spider with a wire. The practical use of the tuning forks was also demonstrated to generate quantitative data by comparing the spider's behavioural response against both the size of the spider and the web.

2. Materials and Methods

2.1. Study Sites

The following localities situated in Southeast Wales were visited for data collection: Allt-yr-yn (51°35′37″ N, 003°00′48″ W. 74 m above sea level), Bargoed Country Park (51°41′00.1″ N 3°13′30.4″ W, 164 m a.s.l.), Cwmcarn Forest (51°38′13.1″ N 3°05′53.3″ W, 193 m a.s.l.), Magor Marsh Reserve (51°34′31.2″ N 2°49′41.7″ W, 5 m a.s.l.) and Wentwood (51°38′31.2″ N 2°48′38.4″ W, 203 m a.s.l.). These sites were chosen based on their accessibility and historic records which indicated the presence of *M. segmentata*. The climatic variables at our studies sites during our survey days (August–September in both years) ranged from a minimum average temperature of 16.9 °C (Allt-yr-yn) to a maximum of 25.9 °C (Magor Marsh) in 2020 and from a minimum average temperature of 17.9 °C (Allt-yr-yn) to a maximum of 21.8 °C (Magor Marsh) in 2021, while the average relative humidity ranged from a minimum of 73.6% (Bargoed Country Park) to a maximum of 80.6% (Allt-yr-yn) in 2020 and a minimum of 71.0% (Cwmcarn Forest) to a maximum of 78.5% (Allt-yr-yn) in 2021.

2.2. Study Species

Metellina segmentata is a small-to-medium-sized orb web spider (adult females are 4–9 mm long) in the Tetragnathidae family [30], which, similarly to its sister species *Metellina mengei* (Blackwall, 1870), construct horizontally inclined orb webs, and is often found waiting for prey in the centre of its hub during the day [31]. *M. segmentata* and *M. mengei* are difficult to differentiate in the field, but as the latter is only active in the spring, we assumed all the adult females tested in this study were *M. segmentata*. A few individuals from each location were also taken to back to the laboratory and confirmed as *M. segmentata*.

2.3. Data Collection

We used two tuning forks (256 Hz and 440 Hz) and an 8 cm length of wire to study the behavioural responses of *M. segmentata*. The frequencies of the tuning forks were chosen to be within the range of previously studied tuning forks, with the expected responses based on observations that smaller prey insects, such as mosquitoes, have higher wingbeat frequencies (500–650 Hz) [32], while larger, and potentially more dangerous insects, such as bees and wasps, have lower wing beat frequencies (100–250 Hz) [33]. We included the wire as we expected the direct physical stimulation to act as a simulation of larger vertebrate predators, and hence potentially to produce a stronger and more consistent anti-predator response. Therefore, our expectations were that 440 Hz would tend to generate predatory responses, while 256 Hz and the wire would generate anti-predatory behaviours.

Data were obtained over two years (2020 and 2021) using slightly different methods. In 2020, data collection took place between the 8th of August and the 6th of September and consisted of using the stimuli in the same order every time; first the 256 Hz tuning fork, followed by the 440 Hz tuning fork, and finally the wire. In 2021, data were collected between the 1st of August and the 17th of August and here we randomized the order in which the stimuli were presented to each spider by using a random number generator prior to going out into the field.

The processes of both methods were the same, with the order in which stimuli were presented to the individual spider being the only difference. After locating the spider in the hub of its web, the tuning forks would be hit on an object (i.e., a plastic spray water bottle) to generate vibrations, and the fork would then be brought slowly towards the web to touch its frame. The wire would touch the spiders directly on the back of its abdomen. Between each stimulus, a minimum waiting period of 5 min and a maximum wait time of 20 min was allowed for the spiders to return to the hub if they had responded by escaping or attacking. Most spiders returned within this time frame. Our methodology was similar to that of Boys [24], Bays [27] and Justice et al. [28].

To quantify the behavioural responses, the entire response of the spider was written down in the field, and used to create an ethogram (Table 1). This ethogram consisted of three broad categories of behaviour, (i) attack, (ii) escape and (iii) no response—these terms will subsequently be used throughout this report.

The spiders were filmed using a Canon EOS 1300D camera (Canon Inc., Ōta, Tokyo Metropolis, Japan) and a Canon EF-S 18–55 mm Macro Lens (Canon Inc., Ōta, Tokyo Metropolis, Japan) —0.25 m/0.8 ft—with a frame rate of 25 fps. Videos S1–S4 recordings allowed us to clarify behavioural responses and enabled us to measure the spiders' reaction to the stimulation. The reaction time was measured from when the stimulus came into contact with the web or spider, to the initiation of the spider's behavioural response. For example, if the spider turned then moved towards the direction of the tuning fork, the reaction time was measured from when the tuning fork came into contact with the web, to when the spider first turned. If the spider dropped from the web before the tuning fork made contact, this resulted in a negative reaction time.

As well as behavioural responses and reaction times, the length of the spider (mm), height (cm) and length (cm) of the capture spiral were measured using a standard 30 cm ruler. The capture spiral area was calculated using the ellipse-hub equation (vertical diameter/2 × horizontal diameter/2 × π—(hub diameter/2)2) [34]. The temperature (°C) and humidity (%rH) were measured using an ETI 810-190 Pocket-Sized Thermo Hygrometer (Electronical Temperature Instruments Ltd., Worthing, U.K.)).

Table 1. Ethogram of anti-predatory and predatory behaviours in *Metellina segmentata*. The ethogram is split into three categories, attack, escape, and no response.

Behavioural Response		Stimulus	No. of Responses
	Attack		
Turning towards the tuning fork	The spider, whilst staying on the central hub, turned to face the direction of the tuning fork, but did not move from the hub.	256 Hz and 440 Hz	2 (2020) 6 (2021)
Moving towards the tuning fork	The spider moved towards the location of the tuning fork, then stopped on the capture spiral, or returned to the central hub.	256 Hz and 440 Hz	23 (2020) 8 (2021)
Touching the tuning fork	The spider moved towards the tuning fork, and then touched it, but remained on the web.	440 Hz	1 (2020) 15 (2021)
Moving towards the tuning fork, then dropping from the web	The spider moved towards the tuning fork, but before touching it/getting close, the spider dropped from the web.	256 Hz and 440 Hz	17 (2020) 15 (2021)
Touching the tuning fork, then dropping from the web	The spider moved towards the tuning fork, after coming into contact with it, the spider dropped from the web.	440 Hz	6 (2020) 2 (2021)
Attacking	The spider moved its legs, to either grab or fight the wire, and remained on the web.	Wire	11 (2020) 12 (2021)
Grabbing	The spider held onto the wire, coming off the web.	Wire	5 (2020) 5 (2021)
	Escape		
Dropping from the web	The spider dropped off the web; either to the floor, vegetation below, or in the air.	256 Hz, 440 Hz and wire	22 (2020) 27 (2021)
Moving/running away	The spider ran away, usually in the opposite direction to the tuning fork, moving off the web onto adjacent vegetation	256 Hz and wire	8 (2020) 8 (2021)
Jumping away	The spider jumped away after being touched by the wire, but remained on the web.	Wire	9 (2020) 10 (2021)
	No response		
Flinching	The spider's body moved slightly. It recoiled, but remained on the web.	Wire	6 (2020) 2 (2021)
No response	The spider did not respond to the stimuli in any way.	256 Hz, 440 Hz, wire	4 (2020) 7 (2021)

2.4. Data Analysis

For the 2020 method, a total of 50 spiders were observed during the initial data collection, and 38 were subsequently used in the data analysis. For the 2021 method, 46 spiders were observed during the initial data collection, and 40 were used in the subsequent data analysis. Spiders were removed due to insufficient data, primarily caused by not returning between stimuli. A Fisher's Exact test for contingency tables [35] was used to compare frequency distribution across the different stimuli and across the years, as well as between the order in which the stimuli were presented, as a Chi-square contingency test could not be used due to many expected values being below 5. We developed linear mixed models (using the *lmer()* function from the lme4 package) in R [36] to test whether the type of behavioural response (using only the data where the spider responded either by attacking or escaping) varied with spider size and capture spiral area, where the latter were response variables and behavioural response a fixed variable. Site and year were included as random variables.

For all models, we log-transformed the response variables to achieve normality of the residuals and p values were estimated from the Type II Wald F tests with Kenward–Roger degrees of freedom. We used R [37] for all statistical tests with a significance level of 0.05.

3. Results

3.1. Behavioural Responses

The produced ethogram shows that our spiders displayed a wide range of responses to the different stimuli, which we could categorise into either attack, escape, or no response behaviours (Table 1). The attack behaviours were the most complex of the three categories and generally involved moving towards the stimuli and in some instances physically interacting with the tuning fork or wire. Interestingly, unlike for escape or no response behaviours, the spiders reacted to the wire with different attack behaviours than they did to the tuning forks. The most common reaction to the tuning fork was an initial aggressive movement towards the tuning fork, but either pausing or dropping from the web before it came into contact with it. The most frequent escape response was to rapidly drop off the web (Table 1). The small proportion of spiders that did not respond mainly showed no visible reaction, although a few flinched slightly (Table 1).

In order to better quantify spiders' reactions, we analysed the behaviours in the larger categories, attack, escape, or no response. We found a significant difference in the responses to the three stimuli for both 2020 (Fisher's Exact Test, $p < 0.0001$) and 2021 (Fisher's Exact Test, $p < 0.0001$), with spiders predominantly escaping in response to the 256 Hz tuning fork, almost exclusively attacking in response to the 440 Hz and either attacking or escaping in response to the wire (Figure 1). The number of spiders not responding remained low for all stimuli, with the highest proportion of 16% found for the wire in 2020 (Table 1).

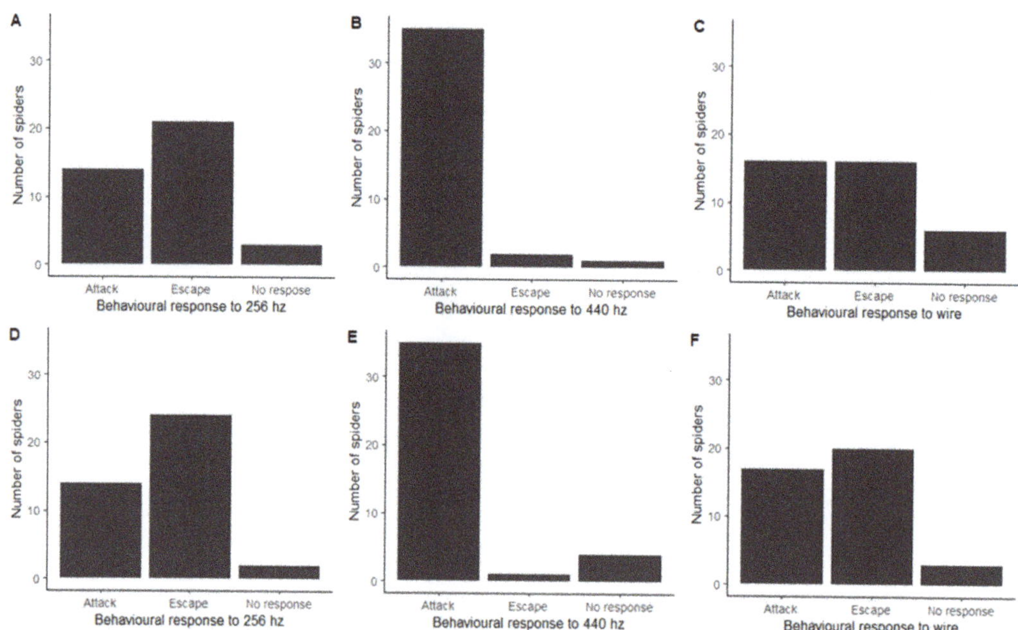

Figure 1. Behavioural responses of *Metellina segmentata*, showing the number of attack, escape, and no response behaviours seen for each stimulus. Split into 2020 method (**A–C**) and 2021 method (**D–F**).

The behavioural responses from the 2020 method and 2021 method were consistent with each other (Figure 1). There were no significant differences in the frequency distributions of responses between 2020 and 2021 for the 256 Hz tuning fork (Fisher's Exact Test,

$p = 0.81$), the 440 Hz tuning fork (Fisher's Exact Test, $p = 0.51$) or the wire (Fisher's Exact Test, $p = 0.49$).

3.2. Order of Stimuli

We found no clear influence of the order of stimuli (which was consistent in 2020 and randomised in 2021), as behavioural responses were similar independently of the order in which they were presented to the spider for both the 256 Hz (Figure 2A–C; Fisher's Exact Test, $p = 0.584$) and the 440 Hz tuning fork (Figure 2D–F; Fisher's Exact Test, $p = 0.333$). However, the behavioural responses to the wire showed less consistency (Figure 2G–I; Fisher's Exact Test, $p = 0.033$). When the wire used as the first stimulus, the primary response was to attack (Figure 2G), whilst when it was the second stimulus, the primary response was to escape (Figure 2H). These results are further confounded by almost equal attack and escape behavioural responses when it was the third stimulus (Figure 2I).

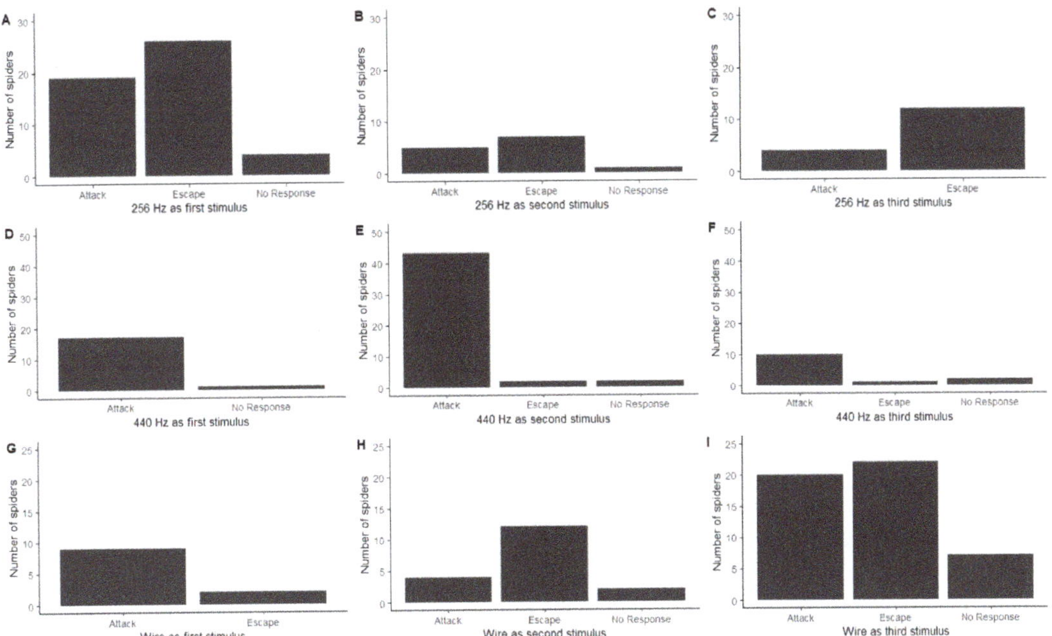

Figure 2. The influence of the order of the stimuli on the behavioural responses of *Metellina segmentata*. (**A–C**) 256 Hz tuning fork; (**D–F**) 440 Hz tuning fork; (**G–I**) length of wire.

3.3. Effect of Size and Web Area

In order to showcase some of the quantitative questions that can be answered using these methods, we compared behavioural responses (attack or escape only) between spiders of different sizes and with different sized webs for the full data-set, disregarding year and order of stimuli as they had no large impact on the behavioural response as demonstrated above. We found that statistically, the length of the spider did not differ between behavioural responses to 256 Hz (F = 2.41, df = 1, 67.5, $p = 0.13$), 440 Hz (F = 1.20, df = 1, 68.5, $p = 0.28$), or the wire (LMM: F = 0.003, df = 1, 63.8, $p = 0.96$). Similar results were obtained with web area, where we also found no differences in behavioural response to 256 Hz (F = 1.22, df = 1, 67.8, $p = 0.27$), 440 Hz (F = 0.47, df = 1, 69.1, $p = 0.49$) or the wire (F = 2.57, df = 1, 63.4, $p = 0.11$). Nonetheless, the average length of spiders and the average capture spiral area tended both to be larger in spiders that attacked (Figure 3). Whilst for the wire, the size of the spider had no impact on the behavioural response observed, although spiders that escaped tended to have a larger capture spiral (Figure 3).

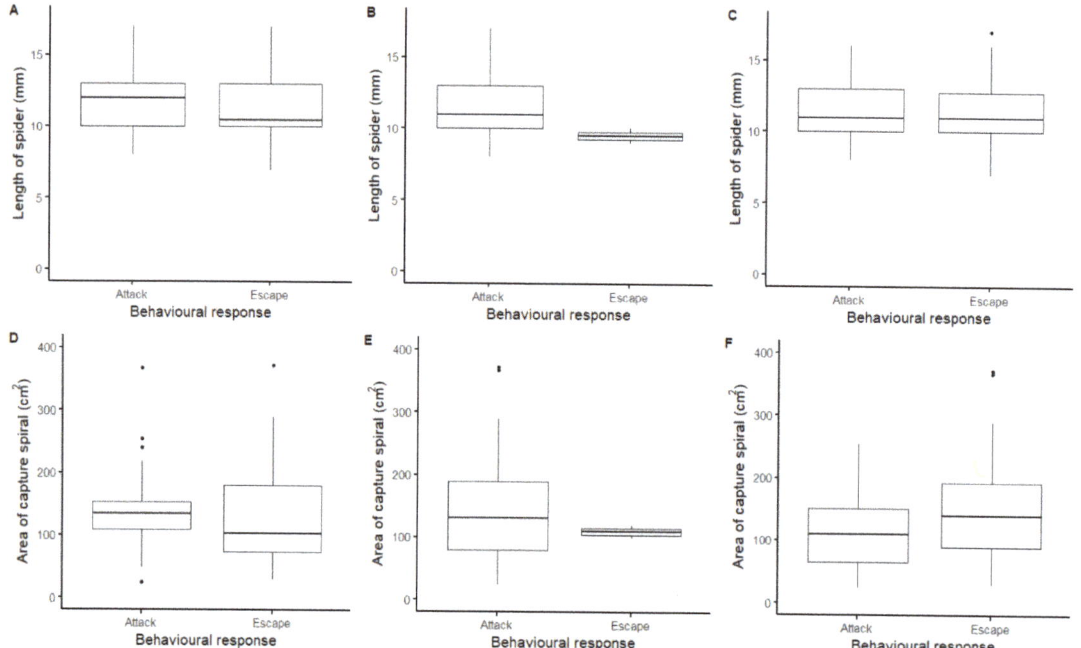

Figure 3. Comparison of the length of the spider (mm) and the area of the capture spiral (cm^2) against the behavioural responses observed in *Metellina segmentata*. (**A**,**D**) 256 Hz tuning fork; (**B**,**E**) 440 Hz tuning fork; (**C**,**F**) length of wire.

3.4. Reaction Times

The stimulus type (tuning forks or wire) and the behavioural response (attack or escape) both had a significant impact on the reaction time of the spider (Behaviour: F = 22.8, df = 1, 202.0, $p < 0.0001$; Stimulus: F = 4.09, df = 2, 200.3, $p = 0.018$) and the interaction between them was also significant (F = 9.47, df = 2, 201.5, $p = 0.0001$). Interestingly, whilst spiders had faster reactions when escaping the 256 Hz tuning fork, they reacted faster when they attacked the 440 Hz tuning (Figure 4). In response to the wire, the difference was minimal, although there was a slight tendency for them to react faster when escaping (Figure 4).

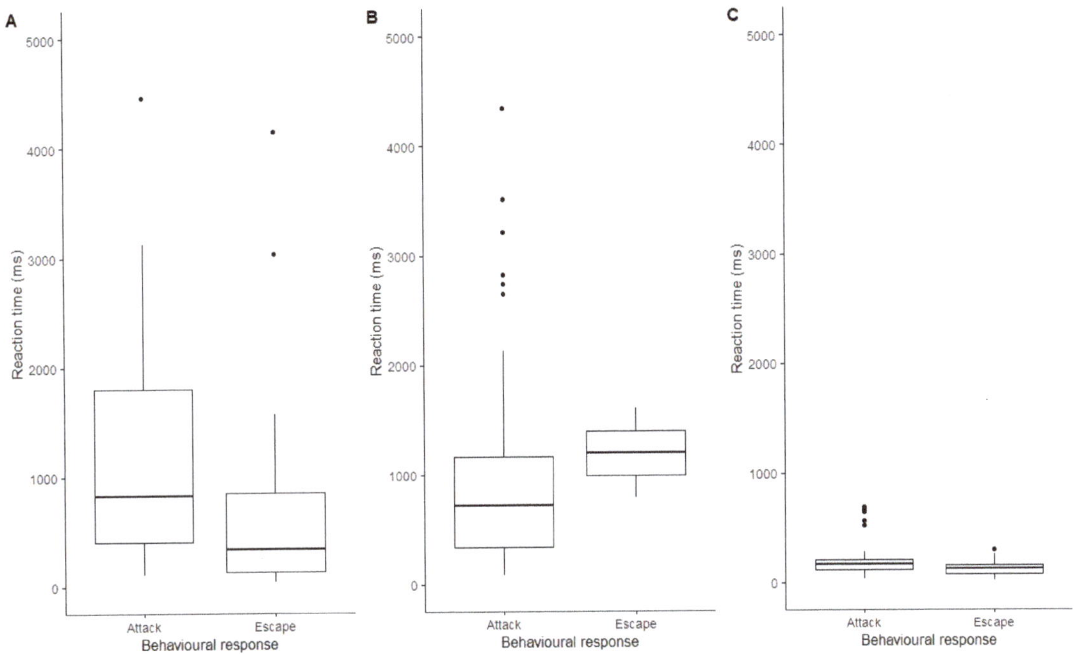

Figure 4. Reaction times (ms) of the attack and escape behavioural responses of *Metellina segmentata* for the three stimuli. (**A**) 256 Hz tuning fork; (**B**) 440 Hz tuning fork; (**C**) Metal wire.

4. Discussion

Our results confirmed that tuning forks can be utilised to study the behavioural responses of orb web spiders. The behaviours demonstrated by *M. segmentata* are in accordance with the secondary defences described by Cloudsley-Thompson [10] and Tolbert [11] in that we observed both rebuff (attacking the stimuli) and flight (running away from the stimuli). The most common escape behaviour in our study was dropping from the web, which was also shown sometimes following attack behaviour. Dropping from the web on a safety thread to enable an easy return to the web after the danger has passed has been described as common in both araneid and tetragnathid spiders [10,38]. In the ethogram developed based on our study, we broadly categorised the behaviours as attack, escape, and no response, with the first group showing the largest number of distinct behaviours. Behaviours in this group also showed a combination of attacking and escaping, with some spiders moving towards the location of the tuning fork, and then escaping by dropping from the web. It should be stated that the spiders observed within this experiment never exhibited the common anti-predatory response of web shaking. Defined by Willey et al. [39] as "violent, large amplitude, movements of the web surface", this behaviour has been observed in *Cyrtophora* [40] and *Mecynogea* [39]. Tolbert [11] called it web flexing, and observed the "Spider and web [swinging] back and forth parallel to the ground ... " in *Argiope* and *Araneus* spp. This behaviour has been observed in the tetragnathid *Azilia vachoni* (Caporiacco, 1954) [38], but it is possible that the Meteine group of tetragnathids do not show this anti-predator behaviour.

The replication of results from three different methods, over two years, demonstrates the feasibility of obtaining reliable behavioural data using tuning forks, and the ease at which they can be used in the field confirm their suitability. The repetition in behavioural responses for the two tuning forks, regardless of the order the stimulus was presented, further demonstrates the reliability and effectiveness of the method. Directly stimulating spiders mechanically with a wire was, while still generating a clearly defined escape or attack behaviour, less reliable. Additionally, the behavioural responses to this stimulus appeared to be dependent on the order of its use; the change from attack to escape when it is the second and third stimuli could be due to the spider becoming sensitised to the prior stimuli. For *M. segmentata*, the 440 Hz tuning fork is the ideal frequency to simulate predators, as the behavioural responses to this frequency are principally attacking (90% of responses for both years combined). The 256 Hz tuning fork largely produces escape behaviours (58% of responses across the two years), although there was more variation with this tuning fork (36% showed attack behaviours). This could mean that a lower frequency is required to produce the ideal simulation of a predator as larger, and hence potentially more dangerous, insects tend to vibrate their wings between 100 Hz and 250 Hz [33]. Alternatively, this inter-individual variation could be attributed to behavioural syndromes, defined by Sih et al. [41] as "a suite of correlated behaviors reflecting between individual consistency in behavior across multiple (two or more) situations". Behavioural syndromes can explain the tendency of some spiders to always attack or always escape from certain stimuli. For example, there are aggression syndromes, where across a range of situations, some individuals are more aggressive, whilst others are less aggressive [41]. The ease of using the tuning fork method in the field mean it could potentially be used to study behavioural syndromes in spiders in the wild.

The size of the spider (here measured as total length) in our study did not significantly affect whether they responded to the stimulation with attack or escape behaviours, although there was a weak trend across the two tuning forks for the attacking spiders being slightly larger than those that escaped. Similarly, the size of the web (here measured as capture spiral area) did not differ between those spiders that attacked or escaped. However, again there was a weak trend for spiders showing attack behaviour in response to the tuning forks to have larger webs, although an opposite trend was found in response to the wire. Our findings are supported by a previous study relating female aggressiveness towards males and prey in the araneid spider *Larinioides sclopetarius* (Clerck, 1757), which found that aggressiveness was largely independent of female size [42].

Our findings further demonstrate how tuning forks can be used to collect quantitative data such as the spider's reaction time to the presented stimuli. Reaction time has been a widely used measure for quantifying foraging behaviour [13,14], and it has also been used to estimate motivation or preparedness of the spider to repair webs under different wind loadings [16]. In our study, spiders showed the fastest reaction time to the wire, and the slowest responses to the 256 Hz tuning fork. However, interestingly the 256 Hz tuning fork elicited a number of spiders to react with escape behaviours before the tuning fork came into contact with the web, thereby producing a negative reaction time. This occurred across the 2020 and 2021 methodologies. This suggests that *Metellina* spiders are able to detect air-borne movements of predator species similar to reactions of airborne vibrations from prey demonstrated in araneid orb spiders [43]. This phenomenon can be explained by trichobothria, which are cuticular filiform hairs that respond to air movements and can be found on the legs and pedipalps of spiders [44,45].

The effectiveness of using tuning forks still needs to be investigated in more detail using more orb spider species and a wider range of frequencies in order to confirm our findings that show that orb web spiders tend to respond to high frequencies with prey capture behaviour, and to low frequencies with anti-predatory behaviours. Given that Nakata and Mori [29] found that both the small-to-medium-sized araneid *Cyclosa argenteoalba* (Bösenberg and Strand, 1906) (similar in size to *M. segmentata*) and the larger araneid *Eriophora sagana* (Tanikawa, 2000) show anti-predator responses to a 440 Hz tuning fork, it is clear that there are family-, environmental- or species-specific factors at play. However, even if the application of the methods in this study should turn out to be limited to Tetragnathid spiders in the Meteine group, the use of tuning forks could still be of potential use in studying the behaviour of Meteine cave spiders [46]. Studying the behaviour of cave animals in general is difficult due to the impediments caused by high sensitivity to disturbances, low population densities and the practical difficulties of observers spending a long time in caves [47]. So, developing easy and quick methods, such as using tuning forks, as demonstrated in this study, is essential. Anti-predatory behaviours are especially interesting to study in caves, as their low nutrient availability results in few species, and hence few predators, which again should reduce any anti-predatory responses [48]. Thus, comparing responses to the 256 Hz tuning fork at different distances into the cave and between spiders with different degrees of behavioural adaptations to the subterranean habitat, such as *Meta menardi* (Latreille, 1804), which modifies its orb web geometry [49], and *Metellina merianae* (Scopoli, 1763), which does not [50], would shed light on the interplay between evolution and phenotypic plasticity (flexibility in web-building behaviour in response to environmental factors) in shaping the behaviours of cave animals.

Supplementary Materials: The following supporting information can be downloaded at: https://www.mdpi.com/article/10.3390/insects13040370/s1, Video S1: *Metellina segmentata* showing attack behaviour in response to a 256 Hz tuning fork. Video S2: *Metellina segmentata* showing escape behaviour in response to a 256 Hz tuning fork. Video S3: *Metellina segmentata* showing attack behaviour in response to a 440 Hz tuning fork. Video S4: *Metellina segmentata* showing attack behaviour in response to direct stimulation by a metal wire.

Author Contributions: Conceptualization M.S.D. and T.H.; methodology M.S.D. and T.H.; data collection M.S.D.; original draft preparation M.S.D.; writing—review and editing T.H.; supervision T.H. All authors have read and agreed to the published version of the manuscript.

Funding: This research received no external funding.

Institutional Review Board Statement: This study did not require any ethical approval.

Data Availability Statement: Link to Figshare will be made available after manuscript acceptance (https://figshare.com/articles/dataset/Response_of_an_orb_spider_to_tuning_forks/19165460).

Acknowledgments: The authors would like to thank Newport City Council, Caerphilly City Council and Woodland Trust for providing permission to research on the sites. MSD would like to thank Raymond Davies for assistance in the field.

Conflicts of Interest: The authors declare no conflict of interest.

References

1. Herberstein, M.E.; Wignall, A.E.; Hebets, E.A.; Schneider, J.M. Dangerous mating systems: Signal complexity, signal content and neural capacity in spiders. *Neurosci. Biobehav. Rev.* **2014**, *46*, 509–518. [CrossRef] [PubMed]
2. Japyassu, H.F.; Laland, K.N. Extended spider cognition. *Anim. Cogn.* **2017**, *20*, 375–395. [CrossRef] [PubMed]
3. Yarger, J.L.; Cherry, B.R.; van der Vaart, A. Uncovering the structure-function relationship in spider silk. *Nat. Rev. Mater.* **2018**, *3*, 18008. [CrossRef]
4. Blamires, S.J. Plasticity in extended phenotypes: Orb web architectural responses to variations in prey parameters. *J. Exp. Biol.* **2010**, *213*, 3207–3212. [CrossRef] [PubMed]
5. Punzo, F. *Spiders: Biology, Ecology, Natural History and Behaviour*; Brill: Leiden, The Netherlands, 2007.
6. Blackledge, T.A.; Kunter, M.; Agnarsson, I. The form and function of spider orb webs: Evolution from silk to ecosystems. *Adv. Insect Physiol.* **2011**, *41*, 175–262. [CrossRef]
7. Zschokke, S.; Herberstein, M.E. Laboratory methods for maintaining and studying web-building spiders. *J. Arachnol.* **2005**, *33*, 205–213. [CrossRef]
8. Hesselberg, T. Exploration behaviour and behavioural flexibility in orb-web spiders: A review. *Curr. Zool.* **2015**, *61*, 313–327. [CrossRef]
9. Scharf, I.; Lubin, Y.; Ovadia, O. Foraging decisions and behavioural flexibility in trap-building predators: A review. *Biol. Rev.* **2011**, *86*, 626–639. [CrossRef]
10. Cloudsley-Thompson, J.L. A review of anti-predator devices of spiders. *Bull. Br. Arachnol. Soc.* **1995**, *10*, 81–96.
11. Tolbert, W.W. Predator avoidance behaviours and web defensive structures in the orb weavers *Argiope aurantia* and *Argiope trifasciata* (Araneae, Araneidae). *Psyche* **1975**, *82*, 29–52. [CrossRef]
12. Uetz, G.W. Foraging strategies of spiders. *Trends Ecol. Evol.* **1992**, *7*, 155–159. [CrossRef]
13. Turner, J.; Vollrath, F.; Hesselberg, T. Wind speed affects prey-catching behaviour in an orb web spider. *Naturwissenschaften* **2011**, *98*, 1063–1067. [CrossRef] [PubMed]

14. Zschokke, S.; Hénaut, Y.; Benjamin, S.P.; García-Ballinas, J.A. Prey-capture strategies in sympatric web-building spiders. *Can. J. Zool.* **2006**, *84*, 964–973. [CrossRef]

15. Mulder, T.; Mortimer, B.; Vollrath, F. Functional flexibility in a spider's orb web. *J. Exp. Biol.* **2020**, *223*, jeb234070. [CrossRef]

16. Tew, E.R.; Adamson, A.; Hesselberg, T. The web repair behaviour of an orb spider. *Anim. Behav.* **2015**, *103*, 137–146. [CrossRef]

17. Watts, J.C.; Herrig, A.; Allen, W.D.; Jones, T.C. Diel patterns of foraging aggression and antipredator behaviour in the trashline orb-weaving spider, *Cyclosa turbinata*. *Anim. Behav.* **2014**, *94*, 79–86. [CrossRef]

18. Blamires, S.J.; Lee, Y.-H.; Chang, C.-M.; Lin, I.-T.; Chen, J.-A.; Lin, T.-Y.; Tso, I.-M. Multiple structures interactively in-fluence prey capture efficiency in spider orb webs. *Anim. Behav.* **2010**, *80*, 947–953. [CrossRef]

19. Tseng, H.-J.; Cheng, R.-C.; Wu, S.-H.; Blamires, S.J.; Tso, I.-M. Trap barricading and decorating by a well-armored sit-and-wait predator: Extra protection or prey attraction? *Behav. Ecol. Sociobiol.* **2011**, *65*, 2351–2359. [CrossRef]

20. Corey, T.B.; Hebets, E.A. Testing the hypothesized antipredator defence function of stridulation in the spiny orb-weaving spider, *Microthena gracilis*. *Anim. Behav.* **2020**, *169*, 103–117. [CrossRef]

21. Foelix, R. *Biology of Spiders*; Oxford University Press: Oxford, UK, 2010.

22. Mortimer, B.; Soler, A.; Wilkins, L.; Vollrath, F. Decoding the locational information in the orb web vibrations of *Araneus diadematus* and *Zygiella x-notata*. *J. R. Soc. Interface* **2019**, *16*, 20190201. [CrossRef]

23. Landolfa, M.A.; Barth, F.G. Vibrations in the orb web of the spider *Nephila calvipes*: Cues for discrimination and orientation. *J. Comp. Physiol. A* **1996**, *179*, 493–508. [CrossRef]

24. Boys, C.V. The Influence of a tuning fork on the garden spider. *Nature* **1880**, *23*, 149–150. [CrossRef]

25. Barrows, W.B. The reactions of an orb-weaving spider, *Eperia sclopetaria* Clerck, to rhythmic vibrations of its web. *Biol. Bull.* **1915**, *29*, 316–332. [CrossRef]

26. Ganske, A.-S.; Uhl, G. The sensory equipment of a spider–A morphological survey of different types of sensillum in both sexes of *Argiope bruennichi* (Araneae, Araneidae). *Arthropod Struct. Dev.* **2018**, *47*, 144–161. [CrossRef]

27. Bays, S.M. A study of the training possibilities of *Araneus diadematus* Cl. *Experientia* **1962**, *18*, 423–424. [CrossRef]

28. Justice, M.J.; Justice, T.C.; Vesci, R.L. Web orientation, stabilimentum structure and predatory behavior of *Argiope florida* Chamberlain & Ivie 1944 (Araneae, Araneidae, Argiopinae). *J. Arachnol.* **2005**, *33*, 82–92.

29. Nakata, K.; Mori, Y. Cost of complex behaviours and its implications in antipredator defence in orb-web spiders. *Anim. Behav.* **2016**, *120*, 115–121. [CrossRef]

30. British Spider Recording Scheme. Summary for *Metellina segmentata* sens. str. (Araneae). 2021. Available online: http://srs.britishspiders.org.uk/portal.php/p/Summary/s/Metellina+segmentata+sens.+str (accessed on 21 November 2021).

31. Tew, N.; Hesselberg, T. Web asymmetry in the tetragnathid orb spider *Metellina mengei* (Blackwell, 1869) is determined by web inclination and web size. *J. Arachnol.* **2018**, *46*, 370–372. [CrossRef]

32. de Nadai, B.L.; Maletzke, A.G.; Corbi, J.J.; Batista, G.E.A.P.A.; Reiskind, M.H. The impact of body size of *Aedes [Stegomyia] aegypti* wingbeat frequency: Implications for mosquito identification. *Med. Vet. Entomol.* **2021**, *35*, 617–624. [CrossRef]

33. Kawakita, S.; Ichikawa, K. Automated classification of bees and hornet using acoustic analysis of their flight sounds. *Apidologie* **2019**, *50*, 71–79. [CrossRef]

34. Herberstein, M.E.; Tso, I.M. Evaluation of formulae to estimate the capture area of orb webs (Araneoidea, Araneae). *J. Arachnol.* **2000**, *28*, 180–184. [CrossRef]

35. Mehta, C.R.; Patel, N.R. Algorithm 643: FEXACT, a FORTRAN subroutine for Fisher's exact test on unordered r x c contingency tables. *ACM Trans. Math. Softw.* **1986**, *12*, 154–161. [CrossRef]

36. Bates, D.; Maechler, M.; Bolker, B.; Walker, S. Fitting Linear Mixed-Effects Models Using lme4. *J. Stat. Softw.* **2015**, *67*, 1–48. [CrossRef]

37. R Core Team. *R: A Language and Environment for Statistical Computing*; R Foundation for Statistical Computing: Vienna, Austria, 2021; Available online: https://www.R-project.org/ (accessed on 3 March 2022).

38. Sewlal, J.N.; Dempewolf, L. Defence mechanisms of the orb-weaving spider *Azilia vachoni* (Araneae:Tetragnathidae). *Living World J. Trinidad Tobago Field Nat. Club* **2005**, *2005*, 50–52.

39. Willey, M.B.; Johnson, M.A.; Adler, P.H. Predatory behaviour of the basilica spider *Mecynogea lemniscata* (Araneae, Araneidae). *Psyche* **1992**, *99*, 153–168. [CrossRef]

40. Lubin, Y.D. The predatory behaviour of *Crytophora* (Araneae: Araneidae). *J. Arachnol.* **1980**, *8*, 159–185.

41. Sih, A.; Bell, A.; Johnson, J.C. Behavioural syndromes: An ecological and evolutionary overview. *Trends Ecol. Evol.* **2004**, *19*, 372–378. [CrossRef]

42. Kralj-Fišer, S.; Schneider, J.M. Individual behavioural consistency and plasticity in an urban spider. *Anim. Behav.* **2012**, *84*, 197–204. [CrossRef]

43. Nakata, K. Spiders use airborne cues to respond to flying insect predators by building orb-web with fewer silk thread and larger silk decorations. *Ethology* **2008**, *114*, 686–692. [CrossRef]

44. Barth, F.G.; Höller, A. Dynamics of arthropod filiform hairs. V. The response of spider trichobothria to natural stimuli. *Philos. Trans. Biol. Sci.* **1999**, *354*, 183–192. [CrossRef]

45. Savory, T. *Arachnida*, 2nd ed.; Academic Press: London, UK, 1977.

46. Hesselberg, T.; Simonsen, D.; Juan, C. Do cave orb spiders show unique behavioural adaptations to subterranean life? A review of the evidence. *Behaviour* **2019**, *156*, 969–996. [CrossRef]

47. Mammola, S.; Lunghi, E.; Bilandžija, H.; Cardoso, P.; Grimm, V.; Schmidt, S.I.; Hesselberg, T.; Martínez, A. Collecting eco-evolutionary data in the data: Impediments to subterranean research and how to overcome them. *Ecol. Evol.* **2021**, *11*, 5911–5926. [CrossRef] [PubMed]
48. Mammola, S. Finding answers in the dark: Caves as models in ecology fifty years after Poulson and White. *Ecography* **2018**, *42*, 1331–1351. [CrossRef]
49. Simonsen, D.; Hesselberg, T. Unique behavioural modifications in the web structure of the cave orb spider *Meta menardi* (Araneae, Tetragnathidae). *Sci. Rep.* **2021**, *11*, 1–10. [CrossRef]
50. Hesselberg, T.; Simonsen, D. A comparison of morphology and web geometry between hypogean and epigean species of *Metellina* orb spiders (family Tetragnathidae). *Subterr. Biol.* **2019**, *32*, 1–13. [CrossRef]

MDPI
St. Alban-Anlage 66
4052 Basel
Switzerland
Tel. +41 61 683 77 34
Fax +41 61 302 89 18
www.mdpi.com

Insects Editorial Office
E-mail: insects@mdpi.com
www.mdpi.com/journal/insects